World Scientific Series on Emerging Technologies – Volume 4

SYSTEM SUSTAINMENT

ACQUISITION AND ENGINEERING PROCESSES FOR THE SUSTAINMENT OF CRITICAL AND LEGACY SYSTEMS

World Scientific Series on Emerging Technologies: Avram Bar-Cohen Memorial Series

Print ISSN: 2737-5862
Online ISSN: 2737-5870

Series Editors: Eran Sher *(Ben-Gurion University of the Negev & Israel Institute of Technology, Israel)*
Jens Rieger *(BASF SE, Advanced Materials & Systems, Research Technology Scouting & Incubation, Germany)*

This compendium provides a comprehensive collection of the emergent applications of big data, machine learning, and artificial intelligence technologies to present day physical sciences ranging from materials theory and imaging to predictive synthesis and automated research. This area of research is among the most rapidly developing in the last several years in areas spanning materials science, chemistry, and condensed matter physics.

Written by world renowned researchers, the compilation of two authoritative volumes provides a distinct summary of the modern advances in instrument — driven data generation and analytics, establishing the links between the big data and predictive theories, and outlining the emerging field of data and physics-driven predictive and autonomous systems.

Published

More information on this series can also be found at
https://www.worldscientific.com/series/wset

World Scientific Series on Emerging Technologies – Volume 4

SYSTEM SUSTAINMENT

ACQUISITION AND ENGINEERING PROCESSES FOR THE SUSTAINMENT OF CRITICAL AND LEGACY SYSTEMS

Peter Sandborn
William Lucyshyn

University of Maryland, USA

World Scientific

NEW JERSEY · LONDON · SINGAPORE · BEIJING · SHANGHAI · HONG KONG · TAIPEI · CHENNAI · TOKYO

Published by

World Scientific Publishing Co. Pte. Ltd.

5 Toh Tuck Link, Singapore 596224

USA office: 27 Warren Street, Suite 401-402, Hackensack, NJ 07601

UK office: 57 Shelton Street, Covent Garden, London WC2H 9HE

Library of Congress Cataloging-in-Publication Data

Names: Sandborn, Peter A., 1959–   author. | Lucyshyn, William, author.

Title: System sustainment : acquisition and engineering processes for the sustainment of critical and legacy systems / Peter Sandborn, William Lucyshyn, University of Maryland, USA.

Description: Singapore ; Hackensack, NJ, USA : World Scientific, [2023] |
    Series: World Scientific series on emerging technologies. Avram Bar-Cohen memorial series, 2737-5862 ; 4 | Includes bibliographical references and index.

Identifiers: LCCN 2022013689 | ISBN 9789811256844 (hardcover) |
    ISBN 9789811256851 (ebook for institutions) | ISBN 9789811256868 (ebook for individuals)

Subjects: LCSH: Systems availability. | System failures (Engineering)--Prevention. | Sustainability.

Classification: LCC TA169 .S26 2023 | DDC 620/.00452--dc23/eng/20220524

LC record available at https://lccn.loc.gov/2022013689

British Library Cataloguing-in-Publication Data

A catalogue record for this book is available from the British Library.

For any available supplementary material, please visit
https://www.worldscientific.com/worldscibooks/10.1142/12860#t=suppl

Desk Editors: Balamurugan Rajendran/Amanda Yun

Typeset by Stallion Press
Email: enquiries@stallionpress.com

Preface

Students in technology disciplines are provided with educations that often focus only on the development, design, and manufacturing of new, "shiny" things. Many of these students rarely, if ever, encounter substantive discussions about the process of taking care of old things. While the consumer may fill their home with the newest technology (phone, car, appliances, televisions, etc.), there is no escaping the reality that their lives are surrounded by, and depend to a great extent on, legacy systems that are old and getting older (e.g., infrastructure, transportation, power generation, etc.) – in this book, we refer to these as "critical systems". The process of maintaining the critical systems that everyone depends on is often overlooked until people's daily lives are impacted, or these system's failures become catastrophic. When critical systems are designed, their sustainment is often not a primary design consideration. Additionally, these systems are rarely adequately resourced for their long-term sustainment, and even when they are, the sustainment budgets are the first thing raided when funds are needed for other more pressing matters.

While there are many books that focus on system reliability, there are fewer treatments of the consequences of system unreliability. The view taken in this book is that systems are more than just hardware and software, in addition they are supply chains, workforces, contracts, business models, acquisition processes and governance. If any of these elements fail, the system may fail and the outcome that the system's stakeholders require may not be achieved. As a result, this book is a mix of engineering, operations research, and policy sciences intended to provide students with an appreciation of the life-cycle costs, consequences, and risks of procuring and sustaining critical systems.

This book is an outgrowth of a mixture of courses developed in engineering and public policy at the University of Maryland that treat the acquisition, life-cycle cost, reliability, maintainability, and supply-chain risks associated with critical systems. Where critical systems are characterized by high procurement costs, long field lives, severe failure consequence, and a general reluctance or inability by their owners/stakeholders to replace them.

This book is intended to be a resource for advanced undergraduate and graduate students in engineering, business, and public policy who want to understand the ramifications of, and processes for, system sustainment. We also use this book as a reference for industry short courses provided to practicing professionals, whom in many cases, were not introduced to system sustainment during their education and are now thrust into the field with minimal preparation.

As an introduction, the book begins with a general description of system sustainment and its relation to other types of sustainment, and an overview of the acquisition process for critical systems. This is followed by three chapters that are focused on determining the demand for parts (i.e., how many parts will be needed to support a system and obtain the desired outcome from the system over its life). These chapters are followed by two chapters that focus on the management of the inventory of those parts and the supply-chain risks associated with obtaining (and continuing to obtain) the required parts. Chapter 8 adds the dimensions of workforce, system capability, cost-benefit analysis, and life extensions to the system sustainment mix. In Chapter 9, the contract realities associated with sustainment are introduced. In addition to these chapters, three appendices are included that introduce analysis "tools" needed to quantitatively analyze system sustainment: discounted cash flow analysis, Monte Carlo analysis, and discrete-event simulation.

The majority of the chapters contain problems of varying levels of difficulty, ranging from alternative numerical values that can be used in the examples included in the chapter text to derivations of relations presented in the text and extensions of the models described. Even for the simple problems, students may have to reproduce (via a spreadsheet or other methods) the examples from the text before attempting the problems. The acronyms and notation (symbols) used in each chapter are summarized in Appendix D. Every attempt has been made to make the notation consistent from chapter to chapter; however, some common symbols have different meanings in different chapters.

The authors are grateful to many people who have made this a much better book with their input. First, we want to thank the hundreds of students who have taken courses at the University of Maryland and seem to somehow always find new and unique questions to ask every time the material is presented. We would also like to thank our numerous colleagues at the University of Maryland and in CALCE, for encouraging the writing of this book.

About the Authors

 **Peter Sandborn** is a Professor in the Department of Mechanical Engineering at the University of Maryland, USA. He is a Director of the Maryland Center of Excellence for Sustainment Sciences (MChESS), a member of the CALCE Electronic Products and Systems Center and the former Director of the Maryland Technology Enterprise Institute (Mtech) at the University of Maryland.

Dr. Sandborn's research group develops life-cycle cost models and business case support for long field life systems. This work includes: system health management, part obsolescence management, counterfeit part management, return on investment models for maintenance planning and outcome-based contract design and optimization. Dr. Sandborn is an Associate Editor of the *IEEE Transactions on Electronics Packaging Manufacturing* and a member of the Board of Directors of the PHM Society. He is the author of over 200 technical publications and several books on electronic packaging and electronic systems cost analysis. He was the winner of the 2004 SOLE Proceedings, the 2006 Eugene L. Grant, the 2017 ASME Kos Ishii-Toshiba, and the 2018 Jacques S. Gansler awards. He is a Fellow of the IEEE, the ASME and the PHM Society. Dr. Sandborn has a B.S. degree in Engineering Physics from the University of Colorado, Boulder, obtained in 1982, and an M.S. degree in

Electrical Science and Ph.D. degree in Electrical Engineering, both from the University of Michigan, Ann Arbor, obtained in 1983 and 1987, respectively.

Willam Lucyshyn is a Research Professor and the Director of Research at the Center for the Governance of Technology and Systems (GoTech), in the School of Public Policy, at the University of Maryland, USA. In this position, he directs research focused on exploring the development, governance, and sustainment of complex critical infrastructure technologies and networks through rigorous interdisciplinary research. His projects have included identifying government sourcing and acquisition best practices and transforming the Department of Defense's sustainment and supply chain management. He has authored over 100 reports and publications and serves on the Editorial Board of *Defense Acquisition Research Journal*. Previously he served as a program manager and the principal technical advisor to the Director of the DARPA. Prior to that, he completed a 25-year career in the U.S. Air Force serving in various operations, staff, and acquisition positions. Mr. Lucyshyn received his bachelor's degree in Engineering Science from the City University of New York in 1971. In 1985, he earned his master's degree in Nuclear Engineering from the Air Force Institute of Technology. He was certified Level III, as an Acquisition Professional in Program Management in 1994.

Contents

Chapter 1

Introduction to Sustainment

"Sustainment" (as commonly defined by industry and government), is comprised of maintenance, support, and upgrade practices that maintain or improve the performance of a system and maximize the availability of goods and services while minimizing their cost and footprint or, more simply, the capacity of a system to endure. System sustainment is a multi-trillion-dollar enterprise, in government (infrastructure and defense) and industry (transportation, industrial controls, data centers, energy generation and others). Systems associated with human safety, the delivery of critical services, important humanitarian and military missions and global economic stability are often compromised by the failure to develop, resource, and implement effective long-term sustainment strategies. System sustainment is unfortunately, an area that has traditionally been dominated by transactional processes with less than optimal strategic planning, policy, or methodological support.

1.1 The Sustainment/Sustainability Landscape

Sustainability and its variants have captured the interest of engineering (and other disciplines) for several decades. Even though sustainability and sustainment are sometimes used interchangeably, these words have unique connotations that depend on the context and discipline within which they are used. The focus of this book is on the sustainment of critical systems, but let us first look at the most prevalent usages of sustainment and sustainability [1.1]:

1

- *Environmental Sustainability* is "the ability of an ecosystem to maintain ecological processes and functions, biological diversity, and productivity over time" [1.2]. The objective of environmental sustainability is to increase energy and material efficiencies, preserve ecosystem integrity, and promote human health and happiness through design, economics, manufacturing, and policy.
- *Economic (Business or Corporate) Sustainability* refers to an increase in productivity (possibly accompanied by a reduction of consumed resources) without any reduction in quality or profitability. Business sustainability is often described as the triple bottom line [1.3]: financial (profit), social (people) and environmental (planet). "Sustainable operations management" integrates profit and efficiency with the stakeholders and resulting environmental impacts [1.4].
- *Social Sustainability* is the ability of a social system to function indefinitely at a defined level of social wellbeing [1.5]. Social sustainability has also been defined as "a process for creating sustainable, successful places that promote wellbeing, by understanding what people need from the places they live and work" [1.6]. Social sustainability is a combination of the physical design of places that people occupy and the design of the social world, i.e., the infrastructure that supports social and cultural life.
- *Technology or System Sustainment* refers to the activities undertaken to: (a) maintain the operation of an existing system (ensure that it can successfully continue to perform and complete its intended purpose), (b) continue to manufacture and field versions of the system that satisfy the original requirements, and (c) manufacture and field revised versions of the system that satisfy evolving requirements [1.7]. The term "sustainment engineering" when applied to technology sustainment activities is the process of assessing and improving a system's ability to be sustained by determining, selecting, and implementing feasible and economically viable alternatives [1.8].

Many specialized uses of sustainability also exist,[1] which overlap into one or more of the categories above, including: urban

---

[1]There are other usages that are not particularly relevant to engineered systems, for example sustainment and sustainability are used as a general programmatic/practice metric; "sustainability" is a term used to refer to what happens

sustainability, sustainable living, sustainable food, sustainable capitalism, sustainable buildings, software sustainment, sustainable supply chains, and many others. Technology and system sustainment, is the topic of this book.

1.2 Defining Sustainment

With so many diverse interests using sustainability/sustainment terminology, sustainment can imply different things to different people. Both sustainment and sustainability are nouns. However, sustainment is the act of sustaining something, i.e., the determination and execution of the actions taken to improve or ensure a system's longevity or survivability; while sustainability is the ability to sustain something or a system's ability to be sustained. Today, *sustain* is defined as keeping a product or system going or to extend its duration [1.10]. The most common modern synonym for *sustain* is *maintain*. *Sustain* and *maintain* may be used interchangeably, however, maintaining most often refers to actions taken to correct or avoid problems, while sustaining is a more general strategic term referring to the management of the evolution of a system. Basiago [1.11] points out that sustainability is closely tied to *futurity*; meaning renewed or continuing existence in a future time. To sustain embraces a philosophy in which the principles of futurity guide current decision-making.

The first use of the word sustainability in the context of man's future was in 1972 [1.12, 1.13], and the term was first used in a United Nations report in 1978 [1.14]. For the history of the origin and development of socio-ecological sustainability, see [1.15, 1.16]. The best-known socio-ecological definition of sustainability (attributed to the "Brundtland Report" [1.17]), is commonly paraphrased as "development that meets the needs of present generations without compromising the ability of future generations to meet their own needs". While the primary context for this definition is environmental (and social) sustainability, it has applicability to other types of sustainability. In the case of technology sustainment, if the word "generations"

---

after initial implementation efforts (or funding ends) where sustainability measures the extent, nature, or impact of adaptations to the interventions or programs after implemented, e.g., in health care [1.9].

is interpreted as the operators, maintainers, and users of the system, then the definition could be used to describe technology sustainment.

At the other end of the spectrum, the U.S. Department of Defense (DoD) defines sustainment as "the provision of logistics and personnel services necessary to maintain and prolong operations through mission accomplishment and redeployment of the force" [1.18]. Sustainment provides the necessary support to operational military entities to enable them to perform their missions. The second, and perhaps more germane defense definition, is in the systems acquisition context. Once a system is developed and deployed the system operations and support phase consists of two major efforts "sustainment and disposal". How do these definitions relate to the design and production of systems? For many types of critical systems (systems that are used to insure the success of safety, mission, and infrastructure critical activities), sustainment must be part of the initial system design (making it an afterthought is a prescription for disaster – see Section 1.3).

In 1992, Kidd [1.15] concluded that "The roots of the term 'sustainability' are so deeply embedded in fundamentally different concepts, each of which has valid claims to validity, that a search for a single definition seems futile". Although Kidd was only focused on socio-ecological sustainability, his statement carries a kernel of truth across the entire scope of disciplines considered in this chapter. Nonetheless, in an attempt to create a general definition of sustainment that would be universally applicable across all disciplines we developed the following. The best short definition of sustainment is the capacity of a system to endure. A potentially better, but longer, definition of sustainment was proposed by Sandborn [1.19]: "development, production, operation, management, and end-of-life of systems that maximizes the availability of goods and services while minimizing their footprint". The general applicability of this definition is embedded in the following terms:

- "footprint" represents any kind of impact that is relevant to the system's customers and/or stakeholders, e.g., cost (economics), resource consumption, energy, environmental, and human health;

- "availability" measures the fraction of time that a product or service is at the right place, supported by the appropriate resources, and in the right operational state when the customer requires it; and
- "customer" is a group of people, i.e., individual, company, geographic region, or general population segment.

This definition is consistent with environmental, social, business, and technology/system sustainment concerns.

1.3 Critical-Systems Sustainment

Critical systems perform safety-, mission-, and infrastructure-critical activities that create the transportation, communications, defense, financial, energy generation, utilities, and public health backbone of society.[2] The cost of the sustainment of these systems can be staggering. For example, the global maintenance, repair and overhaul (MRO) market for airlines is expected to exceed $100B per year by 2026 [1.20]. Amtrak has estimated its capital maintenance backlog (which includes physical infrastructure and electromechanical systems) in the U.S. Northeast Corridor, alone, at around $21 billion [1.21]. The annual cost to operate and maintain the Department of Defense forces (the operating forces budget) was as high as $180 billion in FY 2011 (at the height of overseas operations), but was still at $158 billion in FY 2018 [1.22].

A sustainment-dominated system is defined as a system for which the lifetime footprint significantly exceeds the footprint associated with manufacturing it [1.7].[3] Where "footprint" has the same definition as in Section 1.2. Defining sustainment-dominated systems

---

[2] Another term sometimes used for critical systems is "mission critical".

[3] Although this book focuses on critical systems, there is also a sustainment "tail" for low-cost commercial products as well, namely warranty (see Section 4.4.1.2), and product end-of-life. But, the difference between critical systems and high-volume commercial products is a matter of degree. For critical systems, sustainment may be the dominate life-cycle cost, whereas for commercial products it's contribution may be far less significant.

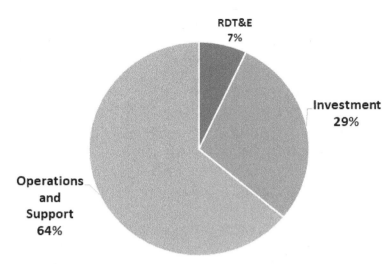

Fig. 1.1.   Average life-cycle cost breakdown of DoD's Fixed Wing Aircraft for an F-16 [1.23]. RDT&E is research, development, test and evaluation.

provides a distinction between high-volume, low-cost consumer products and more complex, higher-cost systems such as airplanes, infrastructure, and industrial equipment. Non-sustainment-dominated products, which tend to be high-volume products, have relatively little investment in sustainment activities and the total time period associated with the product is short (short manufacturing cycle and short field life). Alternatively, sustainment-dominated products, which tend to be relatively low-volume expensive systems, have large sustainment costs and long manufacturing and/or field lives. Some types of sustainment-dominated systems are obvious, e.g., Figure 1.1 shows the average life-cycle cost breakdown for the DoD's fixed-wing military aircraft, where, on average, only 36% of the life-cycle cost of a system is associated with its design, development, test, and manufacturing (this 36% also includes deployment, training, and initial spares). The remaining 64% is for the operation and support, which includes all the costs of operating, maintaining, and supporting, i.e., costs for personnel, consumable and repairable materials, organizational, intermediate, and depot maintenance; facilities, and sustaining investments.

Washington DC Metrorail [1.24]

The Washington DC Metrorail is a 40-year-old system, which is the second busiest subway network in the U.S. after the New York City subway. In the early 2020's Metrorail faced a series of crises. Metrorail needed to manage safety concerns, combat declines in ridership, upgrade its infrastructure, improve its financial management, and find a dedicated stream of funding that allows it to perform more effective strategic planning.

Maintenance on the system that needed to be done has not been done. This is not unusual for transit systems, but it eventually catches up to the system and becomes expensive and frustrating. "This region was not making the investment that WMATA needed going back 15 to 20 years. Now were paying for it", said Chuck Bean, executive director of the Metropolitan Washington Council of Governments.

The Washington DC Metrorail is not alone. In March 2016, a spokesman for the Bay Area Rapid Transit (BART) rail system in the San Francisco bay area, posted the following message to its riders on its Twitter account: "BART was built to transport far fewer people, and much of our system has reached the end of its useful life. This is our reality".

"It's the disconnect between the service we all say we want", and the "political will to pay for something". Christopher Zimmerman, a former Metro board chair.

Sustainment isn't just the most expensive portion of the life cycle for critical systems, it may also be the most profitable part of many critical system programs for contractors. Recently, Lockheed Martin acknowledged that sustainment would become the most profitable part of the F-35 program for it [1.25]: "'Sustainment is going to be the fastest-growing part of the portfolio' as U.S. military services, foreign partners, and foreign military sales customers 'stand up bases' and need steadily increasing numbers of F-35 spare parts, Lockheed CFO Kenneth R. Possenriede said on an April 20 [2021] first quarter results call with stock analysts and reporters. Sustainment will eclipse both F-35 production and development profitability, though he said that development of advanced versions of the fighter is proving a 'pleasant surprise for us,' along with the need to retrofit earlier versions of the aircraft".

Critical systems sustainment is hard because of the confluence of economic realities, discussed in the remainder of this section.

1.3.1   *Technology trends – commercial moves faster than defense*

Prior to and during the Cold War (1947–1991), U.S. technological innovation often occurred as a result of government research funding for military priorities. The resultant technologies (e.g., jet propulsion, satellites, computers, etc.) were then adapted ("spun off") to civilian and commercial applications. With the end of the Cold War, the federal government began to decrease its investment in R&D as a percent of the gross domestic product, from approximately 1.2% to 0.75%, a drop of over 37% (Figure 1.2).

During the same period, as predicted by Gordon Moore in 1965, the number of components per integrated circuit continued to double approximately every 2 years [1.26]. This increased capability of

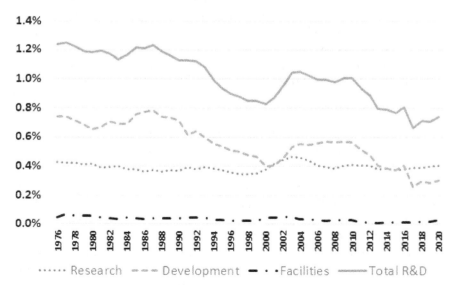

Fig. 1.2.   Federal R&D as a percent of GDP, 1976–2020 (data from American Association for the Advancement of Science).

microelectronics enabled, and continues to enable, a rapid evolution of information and computer technology. As a result, the end of the Cold War coincided with the start of what would become known as "the information revolution".

Research investments, which now increasingly came from the private sector, were focused on purely commercial applications, and the vector of technology transfer had reversed. Government-led programs no longer drove technological innovation as they had in the past; rather, the commercial sector tends to produce much of today's cutting-edge (and, increasingly, globally available) technology. As a result, defense programs frequently lag commercial technology and strive to leverage and adapt the rapidly evolving commercial developments and products. This, of course has many benefits, while reducing development costs. However, at the same time attempting to keep pace with ever changing technologies introduces many challenges. Principal among these is the mismatch between the rapid commercial technological evolution, and the slow-moving government acquisition system and long-lived military systems.

1.3.2 *Globalization of commercial supply chains*

New information technologies facilitate the relatively low-cost and high-speed transfers of large amounts of information. These improvements have reduced transaction costs and the barriers posed by physical borders. Now it is possible to conduct business globally; this has led to the merging of political, cultural, and economic interactions across geopolitical lines. This phenomenon has come to be known as "globalization".

Within these complex networks of global value chains, raw materials can be extracted in one country, exported to a second country for processing, transported to a manufacturing plant in a third country, then shipped to a fourth country for final consumption [1.27]. Consequently, technology development and production, for many products, are globally dispersed. For example, Dell Computers is using this strategy, focusing on global supplier partnerships, mass customization, and just-in-time (JIT) manufacturing, to achieve new levels of efficiency and productivity technology and information, achieving virtual integration [1.28]. Globalization is not limited to the production of goods, but now also includes the provision of services, now growing 60% faster than trade in goods [1.29].

Globalization provides some benefits, including global economic growth; job creation, and making companies more competitive. It also lowers prices for consumer by increasing economies of scale, since the cost per unit produced will decrease as the fixed costs are spread out over a larger number of units produced for a global market. There is also potential for increased competition, as the number of firms available at each stage of the value chain increases. Furthermore, the greater commercial activity between foreign businesses and nations will potentially promote goodwill and cooperation in diplomatic relationships [1.30, 1.31].

However, these advantages are also accompanied with new risks. Primarily there is the prospect of the loss of domestic technical expertise and capability, increasing dependence on global sources and extended supply chains. When these supply chains are interrupted by natural or man-made events, shortages ensue. Additionally, since we have increasingly come to rely on global supply chains, certifying the provenance of parts and components has grown more complex. As a result, the threat of counterfeits has grown in many industries. These parts can potentially affect the safety, operational readiness, and costs of critical systems [1.31].

1.3.3 *Offshoring to China*

Offshoring, that is when firms contract with overseas firms to manufacture parts and components, has grown over the last several decades. This aspect of globalization has increased technology transfer and the nation's dependence on China, a potential peer competitor, and has become an important aspect of its relationship with the U.S.

China has implemented policies to leverage offshoring to domestic capacity while simultaneously closing market opportunities for U.S. companies in important high-tech sectors that include biopharmaceuticals, robotics, and aviation [1.32]. For example, a key objective of the Chinese government has been to the develop its own aerospace industry and technological capabilities. China's ongoing relationship with Boeing, extending back more than 30 years, has played a part in achieving this goal. Every one of Boeing's commercial aircraft incorporates components or modules manufactured in China [1.33]. Roughly, one-third of Boeing's world fleet has major parts and assemblies built in China, with the country having participated in the B787, B777, B747 and B737 projects. For example, China had built the trailing edge wing ribs, vertical fins, horizontal stabilizers, spoilers, and inboard flaps for the B747-8 and the wing panels and doors on the Next-Generation B737 [1.34].

Since 1993, Boeing has also invested in the training and professional development of Chinese aviation professionals by offering free training in piloting, maintenance, flight operations, and management techniques. In November 2012, Boeing launched The Boeing Academy-China to provide more enhanced, integrated training initiatives in China [1.34]. In 2014, Boeing and the state-owned aerospace and defense company, the Aviation Industry Corporation of China (AVIC) collaborated to establish a Manufacturing Innovation Center in one of the AVIC facilities that would provide training for AVIC employees on Boeing's production methods in an effort to improve the employees manufacturing and technological capabilities [1.35].

Indeed, technology transfer is often the result of the intentional sharing of technology and processes. Thus far, this strategy seems to be remarkably successful. AVIC has secured international subcontracts and sub-system joint ventures, largely thanks to cheap labor and large sales prospects, becoming the sole suppliers of some items [1.36]. Specifically, China's extensive involvement in the manufacturing of B737 parts and assembly of A320 for Airbus has improved its knowledge of the development and production single-aisle planes [1.37].

Another example is China's semiconductor manufacturing equipment (SME) industry, which looks in 2020 much like Japan's SME industry of the mid-1970s. Then, Japan's sole strength in SME rested in assembly and packaging tools, benefiting from lower labor costs, just as China does today. But over the next 15 years, Japan's SME industry grew at an astonishing rate, and by the 1990's the global SME market quadrupled in size, and Japanese firms' sales increased eightfold over the same period [1.38].

The growth of China's SME industry ultimately increases its global influence and creates additional vulnerabilities. This was vividly demonstrated during the COVID 19 pandemic. China had become a major global supplier of personal protection equipment, medical devices, antibiotics, and active pharmaceutical ingredients. During the pandemic the Chinese government nationalized control of the production and distribution of medical supplies and directed all production for domestic use. Once they had their internal COVID 19 outbreak under control, the Chinese government prioritized certain countries and selectively released some medical supplies for overseas delivery. The reduction of exports to the U.S. led to shortages of critical medical supplies [1.39].

1.3.4   *Budgetary pressures are forcing organizations to keep older equipment in operation longer*

Government at all levels is facing the prospect of increasing demands for social and medical services, reducing available funds to maintain critical systems and infrastructure. One can look at the U.S. DoD's budget, as an example. Although the budget appears (see Figure 1.3) to be holding above the minimums reached in previous post-conflict periods, the composition of defense spending is changing dramatically. Within the current and projected budgets, the Research, Development, Test, and Evaluation and Procurement accounts are on the

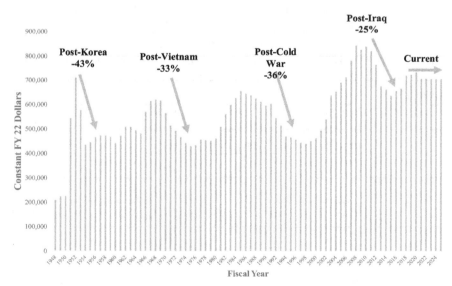

Fig. 1.3.   DOD budgets from 1946–2021, with projections out to 2024 [1.40].

decline, even as the cost of new systems continues to rise. At the same time, personnel and healthcare costs (military retiree healthcare is funded by the DoD) are increasing dramatically [1.41].

As a result, DoD is forced to keep existing systems and equipment in operation longer. This creates a new set of challenges that include increased maintenance costs and obsolescence issues (these often lead to challenges with counterfeit parts), further exacerbating the budget situation. This is commonly referred to as a vicious circle or cycle where a bad situation feeds on itself to get worse. Many "systems" (social, economic, ecological, technological) are comprised of chains of events that form feedback loops that can under some circumstances reinforce the previous cycle (i.e., positive feedback).

It is not uncommon for sustainment-dominated systems to get caught in a "sustainment vicious circle" [1.42]. In this case, more money is going into sustainment (maintaining the systems) at the determent of new investment, which causes the systems to age, which in turn causes more money to be required for sustainment, which leaves less money for new investment, and the cycle continues, e.g., Figure 1.4. The sustainment vicious circle is a reality for militaries in many of the world's countries, civil infrastructure, and can also appear for systems owned by individuals. For example, individuals might face this dilemma with their automobile – fixing your existing car is expensive, but it is less expensive than buying a new car; after several such repairs one is left to wonder if purchasing a new car would have been less expensive, but there is no turning back, too much has been invested in repairing the old car (the funds that could have been used to purchase a new car have been spent).

In this book we are primarily concerned with large, complex, expensive systems, like a fleet of aircraft, a fleet of buses, rail infrastructure, the water/sewer system in cities, etc. These things are massively expensive to replace or upgrade, but also extremely expensive to sustain. Let's be clear, not every large, complex, expensive system suffers the sustainment vicious circle – many are well managed and appropriately funded. However, the evidence of sustainment vicious circles is all around you – look at the neglected portion of a city's aging infrastructure. The cause of the sustainment vicious circle is often the inability or unwillingness of system's owners (which may be the public) to budget appropriately for sustainment (operation

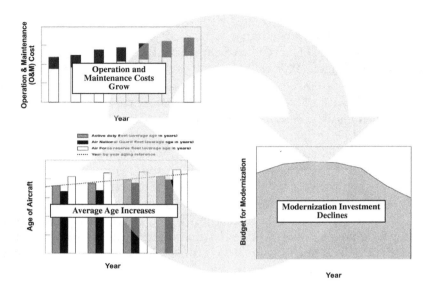

Fig. 1.4.   Sustainment vicious circle for aircraft avionics [1.43].

and maintenance), which is exacerbated by unanticipated life extensions of the system requiring it to last longer than it was designed (or funded) to last. One defining attribute of sustainment-dominated systems is that they always end up having to be supported longer than anyone anticipated.

How do you get out of a sustainment vicious circle? Either the system reaches its breaking point (literally, possibly catastrophically) and simply must be replaced, or the system has to be transitioned to a more "vital cycle" that would reduce the sustainment costs of legacy systems, and provide for modernization. Pursuing the "vital cycle" is not cheap (customers have to be willing to pay) and it usually comes with one key caveat: any action taken cannot adversely affect the customer's needs in the short term.

Related to the vicious circle is the "upgrade trap" [1.7]. A common version of the "upgrade trap" is forced upon consumer electronics users. How often have you been forced to buy a new device because you are receiving files from others that you can't open because the applications needed to open them won't run on your old hardware?

The upgrade trap is indiscriminate, even users that derive no actual benefit from higher-performance hardware or greater software functionality, are forced to "keep up" whether they want to or not. Even systems that are seemingly disconnected from commercial interests such as military systems are impacted too, e.g., if these systems contain Commercial Off-The-Shelf (COTS) application software, the continued support of the application may require hardware upgrades that are not otherwise needed or wanted. Is the upgrade trap by design – certainly, it is a form for planned obsolescence.[4]

The one thing that is worse than being caught in the upgrade trap, is not being caught in the upgrade trap, i.e., being caught in the "sustainment vicious circle".

1.3.5 Shifts from just-in-case to just-in-time (JIT) inventory management

Throughout most of the industrial age firms used "just-in-case" logistics systems. These were dependent on procuring and maintaining large inventories. Although these inventories were often scaled to support anticipated high-demand scenarios and successfully met those needs, they incurred high costs. In addition to the cost of procuring and maintaining the inventories, they also faced obsolescence, as well as disposal costs. Overall, although this created an effective system, it was often far from efficient.

JIT was developed by Japanese auto manufactures in the 1950's in an attempt to eliminate these costs. It is management philosophy that involves having the right items of the right quality and quantity in the right place and the right time. When properly implemented, JIT results in increases in quality, productivity and efficiency, improved communication, and decreases in costs and wastes. After proving its merits in Japan, international firms began to adopt JIT, with its inevitable adoption beginning in the U.S. in the 1970s. JIT allows companies to satisfy orders at lower cost and proved very promising from companies that ranged from McDonalds to Dell. Now instead of selling premade burgers or computers these companies

---

[4]Planned obsolescence is the calculated act of making sure the existing version of a product will become dated or useless within a given time frame.

tailor their products to specific customer needs. Firms were also able to deliberately maintain very low levels of inventory in a further effort to eliminate waste and cost, only ordering enough material and generating enough products to complete orders to meet exact demand [1.44].

However, as firms implemented JIT by reducing inventories and removing redundancies, new vulnerabilities were created. Any disruptions in their supply chains creates a major interruption in their capability to produce their products or maintain their systems. There are many examples, including a 1995 earthquake that disrupted Toyota suppliers' ability to manufacture brake components, Ericsson lost more than $2 billion in sales of cellphones because of a lightning strike, and another earthquake in Japan in 2011 creating multiple significant events when 25% of the world's silicon production was interrupted. As organizations increase their commitment to JIT, their vulnerabilities increase significantly [1.45].

1.3.6   *The complexity of supply chains*

It is generally accepted that supply chains have become more complex in the past several decades. This is due in part to the move to JIT, along with increased globalization and outsourcing. In this context, complex systems are systems made up of numerous components that interact in non-simple ways and often demonstrate emergent properties.

This complexity is not a desirable feature, since it decreases the performance of operations, decreases visibility, complicates decision-making control, and increases the chances of disruptions. Most studies have viewed supply-chain complexity as a multi-faceted, multi-dimensional phenomenon that is driven by several sources that include the large number of suppliers, network complexity, and dynamic complexity.[5] These complex supply chains can interact in unplanned ways and produce unforeseen events making their behavior surprising and unpredictable and, as a result, can be expected to produce more disruptions [1.47].

---

[5]Dynamic complexity is a property of any complex system "where cause and effect are subtle and where the effects over time of interventions are not obvious" [1.46].

Table 1.1. Elements of critical system sustainment.

Affordability	Availability	Policy/Governance	Mission Engineering
Cost-Benefit Analysis	Readiness	System Health Management	Modernization/ Technology Insertion
Warranty	Reliability	Upgradability	Logistics
Maintainability	Obsolescence	Open Systems	Outcome-Based Contracts
Viability	Prognostics	Qualification/ Certification	Sparing
Risk	Testability	Counterfeit Management	
Diagnosability	Workforce	Configuration Control	

1.4 Elements of Critical System Sustainment

While it is easy to map the disciplines listed in Table 1.1 onto managing hardware components and subsystems, sustainment is about more than hardware. Critical systems are composed of combinations of: hardware, software, operational logistics, business models, contract structures, and applicable legislation, policy and governance. If any of these system components fails, the system potentially fails.

1.4.1 *Legacy systems*

A legacy system is a system that is based on, or composed of, methodologies, processes, architectures, technologies, parts, and/or software that is out-of-date (i.e., old). Although software engineering confines legacy systems to software and architecture, in this book we adopt a broader definition of legacy systems that also includes hardware and processes. Since the critical systems we are interested in often have manufacturing and field support lives of 20+ years, for the majority of their life they are legacy systems.

While the term "legacy system" indicates to some that a system is out-of-date, people often do not realize that much of their life and livelihoods depend on legacy systems. In fact, the title of this book could have been "support of legacy systems".

In large part due to the continuing budgetary pressures previously highlighted, legacy systems make up the backbone of much of the world's infrastructure. The decision to keep an old system is most often influenced by economics, i.e., if the customer (which may be the public) is spending hundreds of millions of dollars on a system, they don't want to pay to replace it every couple of years just because the technology it uses has evolved – you may choose to replace your phone every two years, but utilities can't afford to replace a power plant and airlines can't afford to replace their aircraft every 2 years even though better technology becomes available.

1.4.2   *Product life*

There are various types of *lives* that are relevant depending on the goal of the analysis and management of the system. Systems have a *physical life* that extends from the date it is originally fielded to the its end-of-life.[6] The expectation is the support (maintenance) costs will escalate over time. The notion of physical life is static and confined to a single dimension of time with no reference to costs. The *tax life* of a system or asset is the number of years over which the system or asset is depreciated (or amortized). The *economic life* of a system is the period of time over which the system is useful, i.e., the span of time over which the owner or stakeholders care about minimizing the annual cost of owning and operating the system. Generally, book depreciation of an asset occurs over the asset's economic, or useful, life. Generally, the owner/operator of a system should plan to replace the system when it has reached the end of its economic life.

In this book we will primarily concern ourselves with the physical life of systems.

1.4.3   *Logistics*

Logistics has been integral to human survival since mankind began hunting and gathering food. The origin of the word *logistics* is in the French verb *loger* to lodge or to quarter. Its original use was to

---

[6]The required physical life may, and often is, increased after critical systems are fielded. This practice is called "life extension", which complicates system sustainment (see Section 8.7).

describe the movement, supply and maintenance of military forces in the field. In the military, logisticians were historically assigned quartermaster duties and were responsible for: feeding soldiers, providing fodder for horses, procuring uniforms, equipment, weapons, ammunition, and supporting equipment.

The common definition of modern logistics is "the management of the flow of goods, information and other resources, including energy and people, between the point of origin and the point of consumption in order to meet the requirements of consumers". The basic functions that comprise logistics include material flow, distribution, and the life-cycle maintenance and support of a system throughout its intended period of utilization [1.48]. From the systems support perspective, logistics is composed of supply support, test and support equipment, personnel and training, facilities, transportation and material handling, other required resources, and data management.

For many types of systems and in many contexts (e.g., many business schools), the treatment of logistics is confined to a focus on production support: supply-chain management, distribution planning, transportation management, inventory management, and purchasing – these are important factors in manufacturing, but they do not form a complete picture of logistics. In this book we will take a post-manufacturing service view of logistics that is equally limited in its view of logistics and focus on the concepts of sparing, maintainability, availability, mission capability, readiness, and obsolescence.

There is a fine line between logistics and sustainment, and many people use them interchangeably. In [1.18], sustainment is distinguished from logistics, which is "supply, maintenance operations, deployment and distribution, health service support (HSS), logistic services, engineering, and operational contract support". Boeing defines sustainment as a customer need, while logistics are the activities provided to meet that need [1.49]. In other words, logistics is the process of planning and executing sustainment. If sustainment is the capacity of a system to endure (see Section 1.2), then logistics represents the specific activities performed to keep the system going.

1.4.4 *Resilience*

We hear the word "resilience" a lot these days. It's a word that's ripe for misuse and vagueness. There is a general agreement that resilience is the intrinsic ability of a system to resist disturbances.

Another equivalent definition of resilience is the ability to provide required capability in the face of adversity (sometimes referred to as disaster recovery). Resilient design is about managing the ubiquitous uncertainties that constrain current design practices as well as finding ways to overcome an imperfect understanding of system requirements that lead to fragile and ineffectual system designs.

The definitions above are fine, but the real question is what is the "scope" of the system – and this is where views vary. In this book we focus on critical systems, however, the concepts of resilience and resilient design are more prevalent in the building construction, architecture, and communities design space. In the case of sustainment, we are concerned with disruptions (or adversity) from "aging" issues, both technological and non-technological.

When talking about resilient systems, several disciplines think they have a handle on the problem. The control systems folks think that this is their domain, the optimization folks, the PHM (system health management) folks, the reliability engineering folks, sustainability folks, the system engineering folks, machine learning and artificial intelligence folks, as well as others claim primacy. In our experience, none of these disciplines (with the possible exception of some system engineers) really grasps the total scope of resilience. Every discipline wants to gather resilience under their umbrella and grant themselves ownership of the problem space based on their narrow definition of what a system is.

Designing resilient hardware and software (which is the focus of most resilient system design activities) is necessary but not sufficient for creating resilient systems. For a system to be resilient requires:

- reliable (or self-managing) hardware and software;
- a resilient logistics plan (including supply-chain and workforce management);
- a resilient contract structure;
- a resilient business model; and
- resilient governance (policy, laws, and regulations that enable resilience).

One might also include within all of the above aspects, a resilience to changes in the end-of-support for the system. For many systems, the end-of-support is a moving target that is extended ("life extensions") during the use of the system.

This represents a somewhat broader scope than what is generally articulated in the engineering literature, however, in practice, neglecting any of the elements articulated above potentially creates a legacy system with substantial (and potentially untenable) life-cycle support costs.

Resilient design can certainly be applied to other things, e.g., buildings, communities, furniture, information technology, websites, health care, etc. In this broader world one could consider adding the following to the bullets above:

- culturally resilient (does the system transcend cultures and culture changes);
- environmental resilience (i.e., sustainability); and
- resilience to climate change.

Mission assurance is related to resilience. The objective of mission assurance is to create a state of resilience that supports the continuation of an organization's "critical processes and protects its employees, assets, services, and functions. [It] includes the disciplined application of system engineering, risk management, quality, and management principles to achieve success of a design, development, testing, deployment, and operations process" [1.50]. Mission assurance covers the entire design, development, testing, deployment, and operations process. Obviously, mission assurance is broader than any one single system, however, sustainment plays an important role in any mission assurance strategy.

1.4.5 *Performability engineering*

Performability engineering is the evaluation of all aspects of system performance, which encompasses the evaluation of the reliability of systems, its costs, its sustainability, its quality, its safety, its risk, and all of its performance outputs [1.51]. Performability is sustainment plus other system attributes associated with the performance of the system in the field. While some performance measures are included in sustainment, i.e., availability and the other "ilities" in Table 1.1, there are aspects of how the system performs that are application specific (or application-domain specific) that are not included.

1.5   Systems Terminology

There is no perfect (or standardized) terminology to use in a book like this. We also understand that different organizations use different terms and those terms become embedded in their culture. We have elected to use the term "system" to generally refer to the items that this book applies to rather than "product". To us, a "product" implies a commercial item that can be purchased "off the shelf" at a retail location, which does not appropriately describe the types of items that this book is focused on.

A "system", as used in this book, means an integrated combination of elements that are used to perform a specified objective. More fundamentally, the definitions we are using are:

- *Elements*: The hardware and software, but elements also includes the people, facilities, policies, contracts, documents, and databases required to operate and sustain the system.
- *Component* (e.g., part): An independent and replaceable or addressable portion of a system that "fulfills a clear function in the context of a well-defined architecture. A component conforms to and provides the realization of a set of interfaces" [1.52]. A component represents the packaging of a functional-element set that may include hardware and/or software. The difference between a component and a system is that typically a component represents the lowest discrete element that the system designer can access and manipulate; it may also represent the lowest discrete element that can be replaced upon system failure (however, it could be embedded within the lowest discrete element that must be replaced upon system failure). For simple components, designers and sustainers have no ability to modify or repair components, examples: O-ring, resistor, connector, cable. In some cases, the designer may consider a component to be an element embedded within a specific physical object, e.g., a gate (which is addressable by the designer) that is embedded within an FPGA.
- *System*: A group of interacting or interrelated elements. A system is the aggregate of two or more of elements to fulfill a function that

the individual elements alone cannot. Systems may be combined to create more complex systems; in which case the combined lower-level systems may be referred to as subsystems (subsystems generally do not provide a useful function until integrated with other subsystems or systems). Systems can be repaired (via replacement of failed components) upon failure. Examples: automobile, airplane, missile, and power plant.

- *System-of-Systems* (e.g., enterprise systems): "A collection of task-oriented or dedicated systems that pool their resources and capabilities together to create a new, more complex system, which offers more functionality and performance than simply the sum of the constituent systems" [1.53]. A collection of systems becomes an enterprise system when the collection is given a common mission (task, project, outcome). For example, an aircraft and its supporting infrastructure becomes an enterprise when the combination of the aircraft and infrastructure is tasked to perform a function. Measures such as availability are inherently enterprise measures (not system measures). The combined systems that create an enterprise are generally socio-technical networks consisting of human (workforce), hardware, software, policy/regulation, contracting, network infrastructure, supply chains, etc. Examples: a fleet of aircraft, buses or rental cars, a farm of wind turbines, and communication networks.

Each of the above items may have fundamentally different operational and sustainment threats, requiring a set of unique mitigation strategies. Each item has a different time scale, different business environment and different security objectives.

1.6 This Book

The scope of this book is well summarized by the "Gotta Buy a Thing from a Guy" story. Unfortunately, sustainment problems are complex – there is no single master rule or equation to optimize that provides a solution.[7] The taxonomy of the specific topics addressed in

---

[7]Monocausotaxophilia is a term coined by Ernst Pöppel meaning the love of single causes that explain everything. It is human nature to seek simple rules

this book are summarized in Figure 1.5, which also serves as a guild to the organization of this book. We have attempted to organize all the items in Table 1.1 into a network picture that provides some sense of their dependency or connection to one another.

Gotta Buy a Thing from a Guy ...

The following sums up the topic and scope of this book:

Acquisition = How do I buy a thing from a guy?

- Are there requirements that the thing and the guy must meet for me to buy the thing?
- Are the thing and the guy okay?
- What is the process of buying the thing from the guy? (contract/price)

Sustainment = Can I keep buying the thing from the guy as long as I need to?

- Are the thing and the guy going to remain okay as long as I need them? (uncompromised)
- What processes do I put in place to make sure they are okay? (continuous evaluation)
- If the guy goes away, who do I buy the thing from?
- Do I need to find (or create) a second guy to buy the thing from?
- Are there other guys trying to sell me the thing that are not okay?

End of Life = How do I dispose of the thing (and the guy) when I'm done? (reverse logistics)

- How do I make sure others don't get the thing when I'm done with it?
- Is the guy a liability when I don't need the thing anymore?

---

that are straightforward to satisfy, unfortunately there are no simple solutions to system sustainment.

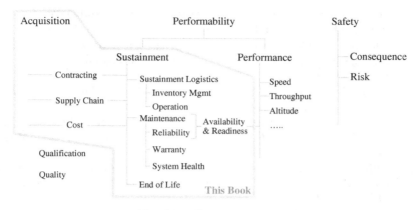

Fig. 1.5.   System taxonomy.

References

[1.1] Sandborn, P. and Lucyshyn, W. (2019). Defining sustainment for engineered systems – A technology and systems view, *ASME Journal of Manufacturing Science and Engineering*, 141(2).

[1.2] Dunster, J. and Dunster, K. (1996). *Dictionary of Natural Resource Management* (UBC Press, Canada).

[1.3] Elkington, J. (1997). *Cannibals With Forks: The Triple Bottom Line of 21st Century Business* (Capstone Publishing, UK).

[1.4] Kleindorfer, P. R., Singhal, K., and Van Wassenhove, L. N. (2005). Sustainable operations management, *Production and Operations Management*, 14(4), pp. 482–492.

[1.5] Social sustainability. thwink.org [cited July 26, 2018]. Available from: http://www.thwink.org/sustain/glossary/SocialSustainability.htm.

[1.6] Wookcraft, S., Bacon, N., Caistor-Arendar, L., and Hackett, T. (2012). Design for social sustainability. Social Life Ltd. [cited July 21, 2018]. Available from: http://www.social-life.co/media/files/DESIGN_FOR_SOCIAL_SUSTAINABILITY_3.pdf.

[1.7] Sandborn, P. and Myers, J. (2008). Designing engineering systems for sustainability, in *Handbook of Performability Engineering*, K. B. Misra (ed.) (Springer, UK), pp. 81–103.

[1.8] Crum, D. (2002). Legacy system sustainment engineering. *Proc. of the DoD Diminishing Manufacturing Sources and Material Shortages*.

[1.9] Wiltsey-Stirman, S., Kimberly, J., Cook, N., Calloway, A., Castro, F., and Charns, M. (2012). The sustainability of new programs and innovations: A review of the empirical literature and recommendations for future research, *Implementation Science*, 7(17), pp. 17–35.

[1.10] Sutton, P. (2004). What is sustainability? *Eingana*, 27(1), pp. 4–9.

[1.11] Basiago, A. D. (1995). Methods of defining "sustainability", *Sustainable Development*, 3(3), pp. 109–119.

[1.12] Goldsmith, E. and Allen, R. (1972). A blueprint for survival, *The Ecologist*, 2(1), pp. 1–43.

[1.13] Meadows, D. H., Meadows, D. L., Randers, J., and Behrens III, W. W. (1972). *The Limits to Growth* (Potomac Associates – Universe Books, USA).

[1.14] United Nations Environment Programme (1978). *Review of the Areas: Environment and Development, and Environment Management.*

[1.15] Kidd, C. V. (1992). The evolution of sustainability, *Journal of Agricultural and Environmental Ethics*, 5(1), pp. 1–26.

[1.16] Du Pisani, J. A. (2006). Sustainable development – Historical roots of the concept, *Environmental Sciences*, 3(2), pp. 83–96.

[1.17] The World Commission on Environment and Development (1987). *Our Common Future* (Oxford University Press, UK).

[1.18] US Government Joint Chiefs of Staff, Joint Publication (JP) 3-0, Joint Operations, January 17, 2017.

[1.19] Sandborn, P. (2017). *Cost Analysis of Electronic Systems*, 2nd Ed. (World Scientific, Singapore).

[1.20] IATA Maintenance Cost Task Force (2017). Airline Maintenance Cost Executive Commentary.

[1.21] Peterman, D. R. and Frittelli, J. (2015). Issues in the reauthorization of Amtrak. *Congressional Research Service*. R42889.

[1.22] Blakeley, K. (2017). Operation and Maintenance. Center for Strategic and Budgetary Assessments. October 31, 2017. Available at: https://csbaonline.org/reports/operation-and-maintenance.

[1.23] Cost Assessment and Program Evaluation (CAPE) (2020). *Operating and Support Cost-Estimating Guide*. Office of the Sec. of Defense. September 2020. Available from: https://www.cape.osd.mil/files/OS_Guide_Sept_2020.pdf.

[1.24] Vack, D. C. (2016). "Fixing WMATA, the Nation's Second Busiest Transit System, From Every Direction", *e.Republic*, October 12. Available at: https://www.governing.com/archive/gov-wmata-transit-problems.html.

[1.25] Tirpak, J. A. (2021). Sustainment becoming most profitable part of F-35 for Lockheed Martin, *Air Force Magazine*, April 20.

[1.26] Moore, G. E. (1965). Cramming more components onto integrated circuits, *Electronics*, 38(8), pp. 114–117, April 19.

[1.27] UNCTAD. (2013). Global Value Chains and Development. United Nations Conference on Trade and Development. 2013. Available at: https://unctad.org/system/files/official-document/diae2013d1_en.pdf.

[1.28] Magretta, J. (1998). The power of virtual integration: An interview with Dell Computer's Michael Dell, *Harvard Business Review*, 76(2), pp. 72–84.

[1.29] Lund, S., Manyika, J., Woetzel, J., Bughin, J., Krishnan, M, Seong, J., and Muir, M. (2019). Globalization in Transition: The Future of Trade and Value Chains. McKinsey Global Institute. January 2019. Available at: https://www.mckinsey.com/~/media/McKinsey/Fea tured%20Insights/Innovation/Globalization%20in%20transition%20 The%20future%20of%20trade%20and%20value%20chains/MGI-Glo balization%20in%20transition-The-future-of-trade-and-value-chains-Full-report.ashx.

[1.30] Collins, M. (2015). The pros and cons of globalization, *Forbes*, May 6.

[1.31] Gansler, J. S., Lucyshyn, W., and Rigilano, J. (2013). The Impact of Globalization on The U.S. Defense Industrial Base. Naval Postgraduate School. July 2013. Available at: https://dair.nps.edu/bitstream/123456789/2597/1/UMD-AM-13-101.pdf.

[1.32] Koleski, K. (2017). The 13th Five-Year Plan. U.S.-China Economic and Security Review Commission, Staff Research Report (February 14).

[1.33] Crane, K., Luoto, J., Warren Harold, S., Yang, D., Berkowitz, S. K., and Wang, X. (2014). The Effectiveness of China's Industrial Policies in Commercial Aviation Manufacturing. Rand Corporation. 2014. Available at: https://www.rand.org/pubs/research_reports/RR245.html.

[1.34] Wang, Y. (2015). Boeing in China Backgrounder. Boeing. Available at www.boeing.com/resources/.../china/.../boeing-china-back grounder-q32015_en.docx.

[1.35] Van Wagenen, J. (2015). Boeing inks partnerships to further China's aviation efforts, *Aviation Today*, September 24.

[1.36] Goldstein, A. (2006). The political economy of industrial policy in China: The case of aircraft manufacturing, *Journal of Chinese Economic and Business Studies*, 4(3), pp. 259–273.

[1.37] Bradsher, K. (2017). China's new jetliner, the Comac C919, takes flight for first time. *New York Times*. May 5.

[1.38] Hunt, W., Khan, S. M., and Peterson, D. (2021). China's Progress in Semiconductor Manufacturing Equipment. Center for Security and

Emerging Technology. March. Available at: https://cset.georgetown.
edu/wp-content/uploads/CSET-Chinas-Progress-in-Semiconductor-
Manufacturing-Equipment.pdf.

[1.39]   CRS. (2020). COVID-19: China Medical Supply Chains and Broader
Trade Issues. Congressional Research Service. December 23. Available at: https://crsreports.congress.gov/product/pdf/R/R46304.

[1.40]   OUSD (Comptroller). (2020). National Defense Budget Estimates
or FY 2021. April 2020 Available at: https://comptroller.defense.
gov/Portals/45/Documents/defbudget/fy2021/FY21_Green_Book.
pdf. Gansler, J. S., Lucyshyn, W., and Rigilano, J. (2013). The
Impact of Globalization on The U.S. Defense Industrial Base, Naval
Postgraduate School Report, July 2013.

[1.41]   CBO. (2016). Long-Term Implications of the 2021 Future
Years Defense Program. Congressional Budget Office. September 2020. Available at: https://www.cbo.gov/publication/56554#_
idTextAnchor001.

[1.42]   Achieving an Innovative Support Structure for 21st Century, PSB
96 (10-13-97).

[1.43]   Ardis, B. (2001). Viable/affordable combat avionics (vca) implementation update. *Dayton Aerospace*, Inc., June 2001.

[1.44]   Kootanaee, A. J., Babu K. N., and Talari, H. F. (2013). Just-in-time
manufacturing system: From introduction to implement, *International Journal of Economics, Business and Finance*, 1(2), pp. 7–25.

[1.45]   Michelsen, C. J., O'Connor, P., and Wiseman, T. (2014). Just in
Time. Expecting Failure: Do JIT Principles Run Counter to DoD's
Business Nature? *USA Defense AT&L*: March–April, pp. 32–38.

[1.46]   Senge, P. M. (2006). *The Fifth Discipline: The Art and Practice of
the Learning Organization*, 2nd Ed. (Doubleday, USA).

[1.47]   Bodea, C. and Wagner, S. M. (2012). Structural drivers of upstream
supply chain complexity and the frequency of supply chain disruptions, *Journal of Operations Management*, 26, pp. 215–228.

[1.48]   Blanchard, B. S. (1992). *Logistics Engineering and Management*, 4th
Ed. (Prentice Hall, USA).

[1.49]   Logistics Support Systems: A New Name for a New Future, *Boeing
Frontiers*, 4(3), July 2005, Retrieved on April 10, 2022. https://www.
boeing.com/news/frontiers/archive/2005/july/i_ids1.html.

[1.50]   Grimm, J. (2004). The Role of CMMI in Mission Assurance,
November 16, 2004. Retrieved on April 21, 2022. https://ndia
storage.blob.core.usgovcloudapi.net/ndia/2004/cmmi/CMMIGS/
NDIARoleofCMMIinMA final.pdf.

[1.51] Misra, K. B. (2008). *Handbook of Performability Engineering* (Springer-Verlag, UK).

[1.52] Kaisler, S. H. (2006). *Software Paradigms* (Wiley, USA).

[1.53] Mistrik, I., Galster, M., Maxim, B. R., and Tekinerdogan, B. (eds.) (2021). Knowledge Management in the Development of Data-Intensive Systems (CRC Press, USA).

Chapter 2

The Acquisition of Critical Systems

Organizations that operate critical systems use the system's readiness or operational availability (A_o) as a key indicator of its success. These metrics are usually expressed as the percentage of total units available and capable of performing a mission at any given time. If a system is not ready when it is needed, its performance characteristics are of no use. There are two ways to achieve high readiness. One approach is to support the system with an extensive logistics system that can provide the required spare parts and other support when and where they are needed. An alternative approach is to design and develop systems that are highly reliable, thus minimizing the support required. The mechanism most generally used to manage the trade-offs between these two approaches is to calculate the system's total ownership costs (TOC), i.e., the cost to develop, produce, operate, and maintain the system through its life cycle. The goal then is to design new systems to be more reliable (fewer failures) and more maintainable (fewer resources needed) without excessive increases in the cost of the system or spares. The result will be a system that meets the necessary system readiness, with a lower TOC.

The remainder of this chapter will review the concept of TOC, a model for the life-cycle acquisition process, system design for sustainment, and what have been identified as industry best practices. We also include specific processes and examples from the public sector, which is often directly or indirectly involved with the acquisition of critical systems. A quantitative treatment of availability and readiness is covered in Chapter 5.

2.1 Total Ownership Costs – Performance Trade-offs

A key objective when developing a system is to ensure its design meets its performance requirements. However, at the same time, affordability over the system's life-cycle is emerging as a critical companion objective. Since many critical systems are operational for 30 years or more, the operating and support costs are typically the highest portion of a system's TOC – they represent the cost to operate the system and maintain its readiness over many years. The Department of Defense (DoD), which develops many critical systems with extended lives, estimates these costs to be 72% of the TOC (see Figure 2.1). This change in focus will require new system engineering practices, trade-off analyses, and other modifications to the system engineering methodologies then have been employed in a performance-first focus. This may require some realignment of organizations and their priorities to ensure that the emphasis on TOC receives the same priority compared to design for performance.

As depicted in Figure 2.1, most of a system's TOC is made up of its operating and support cost, which are incurred after the system is fielded. However, most of the decisions that affect these costs are made early in the acquisition process, during the system's

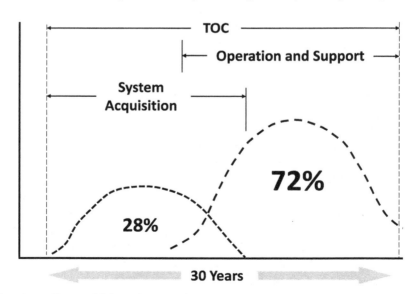

Fig. 2.1.   Nominal life-cycle cost of typical DoD program with a 30-year service life (adapted from the DAU).

development. As shown in Figure 2.2, DoD studies have shown that approximately 85% of the operating and support costs of a weapon system will be determined as soon as the system requirements are finalized. At that point, less than 10% of the life-cycle cost (LCC) has been spent [2.1].

To develop systems that meet the requirements in terms of performance, sustainability, and TOC, a necessary condition is a structured product development process that can make the necessary trade-offs to satisfy the often-conflicting constraints. The process itself will not, however, ensure that the appropriate decisions are made. For many systems, sustainment is an afterthought, and once systems are deployed, resource constraints often preclude replacing them.[1] Because so much of the eventual costs to support and maintain these

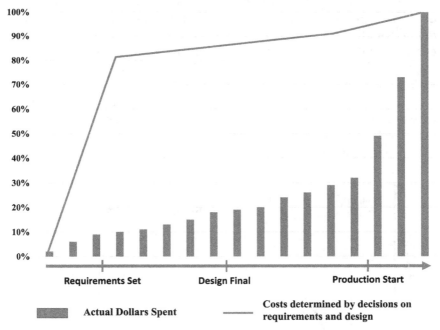

Fig. 2.2. The TOC is set early in a system's life cycle (adapted from the DAU).

---

[1]In 2016 the Internal Revenue Service, for example, was still using main-frame computers and software developed in a programming language few people understand [2.2]. Sustaining these old systems is expensive, and their failure can have significant impacts.

critical systems are decided early in their life cycles, it makes sense that more attention should be paid to their sustainment when the system's requirements are being set, and its design is being finalized. World-class commercial companies that either use or develop high-performing systems know this and set requirements and develop designs accordingly.

2.2   The Acquisition Process

Many firms use some variation of the Project Management Institute's project management procedure (shown in Figure 2.3) to develop new systems. The entire procedure is composed of the five processes described briefly below.

The initiating process are the steps taken to define a new project or a new phase of an existing project where the initial scope is defined and initial financial resources are committed. Internal and external stakeholders will interact to help determine the operational requirements. The principal aim during this phase is to align the stakeholders' expectations for the new system and give them visibility into the scope and objectives. During the planning processes for a new system, the objectives are refined, and the engineering designs are developed. During the executing processes, the system moves into production. Monitoring and controlling processes track, review,

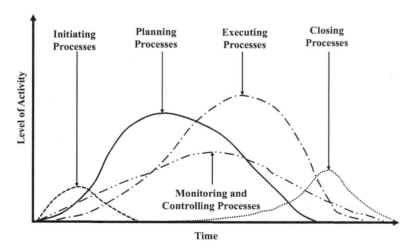

Fig. 2.3.   Project Management Institute's project management process [2.3].

and orchestrate the progress and performance of the project; identify any areas in which changes to the plan are required and initiate the corresponding changes.

The Case of the Washington Metro

The Washington Metro is a rapid transit system serving the Washington DC metropolitan area and began operations in 1976. Metro serves the District of Columbia, as well as several jurisdictions in the states of Maryland and Virginia, and today, operates 91 stations and has 117 miles of track [2.4]. It is the second-busiest rapid transit system in the United States, in number of passenger trips [2.5].

The system has been plagued with safety and reliability problems. According to its own data, only 69% of customers arrive on time, and more than one in four trips are delayed. A Metro rider commuting to and from work five times a week would, on average, experience one delay each week of more than 5 minutes [2.6]. In contrast, the Tokyo subway network is bigger and much more complicated and is claimed to be the best and most reliable. It is normal to see people fast asleep while riding, waking to an alarm set on their phones when they arrive at their station, at the scheduled time [2.7].

System failures have led to several fatal Metro accidents. Among these is the June 22, 2009, collision of two Metrorail trains that resulted in the death of eight passengers and a train operator and injuring 52 others. In January 2015, an electrical arcing incident near an underground station produced toxic smoke, resulting in one passenger fatality and 90 injuries. This incident and another similar fire triggered a system-wide shut down that lasted 29 hours in March 2016.

One series of railcars, the 4000 series, were notoriously unreliable, with more door, propulsion, brake, and air conditioning issues than any other series in Metro's fleet of rail cars. They averaged a breakdown every 26,000 miles, causing twice as many issues as the next lowest performers (worse than the older

(Continued)

(Continued)

series 1000, 2000, and 3000, and the newer 5000 and 6000). Although these cars only made up a small percentage of the fleet, the system is only as reliable as its least reliable cars. Even if there was one set of 4000 series cars in operation on one line, if it were to break down, it could stop all the trains behind it, and possibly even those going in the other direction [2.8].

For their newest acquisition of the Series 8000 railcar, WMATA specified a desired fleet reliability of 200,000 miles Mean Distance Between Delay (MDBD). A delay is defined as an incident that causes a revenue train to be offloaded, a 4 minute or greater delay in service, or a lost trip [2.9]. This is almost an order of magnitude improvement over the troubled series 4000 railcars.

The closing processes in Figure 2.3 are performed to formally complete the project, phase, or contractual obligations. In the case of system development, this would include delivery and, if required, product sustainment. There are often project review points, or gates, between project phases that allow managers to evaluate progress and determine whether to proceed with the development.

The extremely high reliability required from mission-critical systems necessitates accelerated testing, with feedback, to improve designs. When high system reliability is not achieved, systems perform poorly and even cause a loss of life. For example, the Washington D.C. Metro, managed by the Washington Metropolitan Area Transit Authority (WMATA), has been characterized as having the worst reliability of any mass transit system in the world (see the Washington Metro inset).

Leading commercial companies also consider the TOC of new products when making development decisions. This is especially true of firms that acquire long-lived systems such as commercial airlines and tanker ships. These systems are required to operate safely, reliably, and at as low an operating cost as possible, to increase revenues and profits. As a result, the systems may be required to be available virtually 100% of the time and operate at the lowest cost possible.

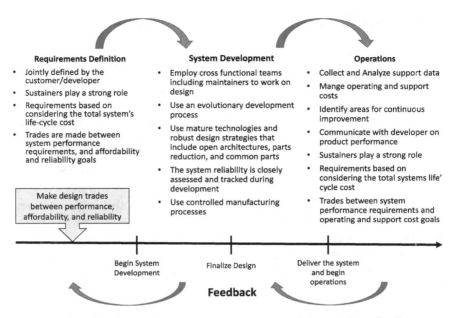

Fig. 2.4. Collaborative development process (adapted from [2.1]).

Consequently, considering reliability and sustainment in the initial design mitigates the need for a retrofit or redesign to improve readiness when the system is fielded, when these options may be more limited. The most critical decisions are made early in the development process, and when setting the requirements. See the Boeing example in the inset.

Consequently, many of these firms work collaboratively with the system developers to help set the system's requirements, focusing on affordable and reliable system sustainment, as well as the system's performance (see Figure 2.4). There is an understanding that additional investment may need to be made upfront (resulting in a higher initial cost) to achieve results that will achieve a lower TOC. When the requirements are established the system developer can focus on these design constraints to finalize the design. When the system is fielded the organizations that operate and maintain the system collect data and provide feedback to the developers to inform modifications and future designs.

2.3   Government Critical Systems Acquisition

Government is often involved, either directly or indirectly, in the acquisition of critical systems. The DoD has, perhaps, the most well documented, publicly available acquisition process within the federal government. It has been reformed numerous times in an effort to improve its performance. However, it continues to produce systems that may not be as sustainable as required, the flaw is generally with its implementation, not its structure.

The terms *procurement* and *acquisition* are often considered synonymous. Within the DoD, however, the terms have different connotations. Acquisition implies much more than just purchasing a system or service; the acquisition process encompasses the design, engineering, construction, testing, deployment, sustainment, and disposal of systems. Whereas the term "procurement", refers to the act of buying goods and services, and it is one of the functions that's part of the acquisition process (e.g., office products are procured).

Boeing Builds Sustainability into the Design [2.1]

When Boeing was developing the Boeing 777, a wide-body airliner, United Airlines was the launch customer and they established stringent requirements for aircraft readiness and operating costs; this ensured reliability would be an important design element. United wanted a twin-engine airplane that could fly extended ranges from any airport in the U.S. and required that the Boeing 777 be available for departure within 15 minutes of the scheduled departure 98.5% of the time.

Boeing guaranteed the aircraft would meet the departure requirement by the third year of operation or pay a penalty – the requirement was achieved by the third year. Boeing agreed to reimburse United for lost revenues when airplanes were unavailable. Boeing also brought together a working group of other potential customers to help develop requirements for the new design. A critical consideration during those meetings were the operating and maintenance costs and projected life-cycle cost, which were used to shape the requirements for the aircraft.

(Continued)

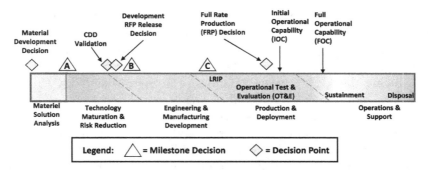

Fig. 2.5. The DoD process for acquisition of hardware-intensive systems.

(*Continued*)

A senior mechanic, who had previously worked for United, was added to the working group to help define the maintenance requirements the ensure aircraft would be easy for mechanics to repair. Boeing implemented an open-systems concept to reduce the total ownership cost of the 777 by allowing customers to choose from three different types of engines – GE, Pratt & Whitney, or Rolls Royce – depending on their needs. Boeing has found that involving the customer in early design decisions improves their ability to design an aircraft that reduces the life-cycle operating cost and is easy and less costly to maintain. That is, it helped ensure that Boeing would build sustainability into the design.

In addition to the Federal Acquisition Regulations (FAR), the Defense Acquisition System is also governed by DoD Directive 5000.01 and DoD Instruction 5000.02. DoD has tailored processes for hardware-intensive systems, software-intensive systems, business system, and services. For the purposes of this book, we will focus on a hardware-intensive model. This process is made up of five phases: the Materiel Solution Analysis (MSA) phase, Technology Maturation and Risk Reduction, the Engineering and Manufacturing phase, Production and Deployment, and finally Operations and Support (see Figure 2.5). Milestones are represented in this figure by triangles placed at the top of the diagram. The milestones are decision points that must be satisfactorily passed before the program can proceed to the next phase of the acquisition process. There are specific statutory and regulatory requirements that a program must meet to proceed to the next phase of the acquisition process.

During the MSA phase, the most promising technology that can meet the requirement is selected. A significant activity during MSA is to complete the Analysis of Alternatives to assess a range of potential materiel solutions that could satisfy validated requirements and identify the most cost-effective solution. Additionally, the effort to secure necessary funding is initiated, and an initial LCC Estimate for the program is submitted. The initiative then undergoes a Milestone A review. It is during this phase that the greatest flexibility in defining the system's sustainment strategy exists. The fundamental goal is aligning broad sustainment requirements with the user's requirements. The initiation of system development is the most opportune time to advocate for standardized systems, components, spare parts, and support equipment.

Once Milestone A has been successfully completed, the program is officially established. The program then enters the second phase, Technology Maturation and Risk Reduction. During this stage the best solution is brought forward and begins its maturation process. During this phase, competitive prototypes may be developed, cost-performance trade-off analyses are conducted, and the Capability Development Document (CDD), which describes measurable and testable capabilities to guide the EMD phase, is prepared. The objective during this phase is to reduce the technology, engineering, integration, and LCC risk to the point that a decision to contract for the Engineering and Manufacturing Development (EMD) phase can be made with confidence in a successful program execution for development, production, and sustainment. It is during this phase that the sustainment design features to achieve the user requirements are incorporated in the overall design specifications.

When the technology maturation and risk reduction are completed, the decision to commit resources and release the solicitation (a Request for Proposals) to industry, is made. For this decision, all the plans for the program are carefully reviewed to ensure all risks are understood and under control, the program plan is sound, and the program will be affordable and executable. Now the program can undergo a Milestone B review, and the program can transition to EMD, and the award of a contract is authorized. The award of the development contract is a critical point in an acquisition program; it is then that resources are committed to a specific system, its performance, and development and production schedule.

During the EMD Phase the system is designed and developed, all technologies and capabilities are fully integrated into a single system, and preparations are made for manufacturing (including developing manufacturing processes, designing for mass production, and managing cost). Additionally, operational testing and evaluation is also conducted at the subsystem, as well as the integrated-system level. The purpose of this testing is to determine whether a system is operationally effective, suitable, and survivable. From a sustainment perspective the objective during this phase is to ensure that an integrated sustainment system that meets readiness targets, is able to meet system performance capability threshold criteria, manages operations and support cost, and optimizes the sustainment footprint is developed. Specifically, this is where supportability design features are incorporated into the maturing design, after trade-offs with performance, size, and LCC.

When the system's design is stable, the Milestone C decision on whether to proceed to production, is made. For this milestone, the system must also pass developmental testing and an operational assessment, and the program must demonstrate that the system is interoperable with other relevant systems and that it can be operationally sustained. It is at this Milestone that the decision to obligate the resources sufficient to enter production and begin deployment of the system is made. The commitment to enter production is generally very expensive and difficult to reverse.

Low-rate initial production is now authorized. The objective is to make ready the manufacturing and quality control processes for a higher rate of production, as well as provide production models for operational test and evaluation. Additionally, during the initial deployment, the sustainment supply chain performance is closely monitored to identify any opportunities for improvement, since this is the first real test it has faced. Once a program has completed the appropriate operational testing and evaluation and demonstrated satisfactory processes, the program will receive approval to enter full-rate production.

The final phase in the acquisition process is the Operations and Support phase. During this phase, the system is fully deployed, operated, supported, and ultimately retired. As previously stated, up to 70% of the TOCs of a system occur during this phase. The program's sustainment strategy must be continually reviewed and assessed. Now, changes to system sustainment are driven by reliability, obsolescence, and other maintenance issues, although changes to the operational and mission requirements can require changes to system sustainment strategy.

2.4 System Design for Sustainment

The most effective actions to reduce sustainment costs are those that are taken during the system's development. As a result, sustainment must play a significant part in a system's initial requirements and design process to affordably achieve the desired sustainment outcomes. These requirements include the availability, reliability, and maintainability necessary to meet the operational needs. When these have been determined, allocations can be made to key subsystems, and these can be tracked during the development process. It is at this time in the life of the system that the trade-space for having the most significant impact on its sustainment is the largest; the objective should be to develop a design that balances the performance and cost over the life of the system. To have the most significant impact, planning for sustainment must begin from the system's initiation and set the appropriate requirements for availability, reliability, and maintainability. Other strategies may be used with mission-critical systems that include shared redundancy where one subsystem may compensate for the failure with another subsystem to enable temporary degraded operation [2.10].

2.4.1 *Setting requirements: Availability, reliability, and maintainability*

System availability is generally defined as the probability that the system will be operational at a random instant in time. We can define a theoretical availability for systems that can be repaired, which assumes that there are unlimited spares and no delays in the provision of maintenance. This value is known as the system's inherent availability or A_i and reflects the system's reliability and maintainability (R&M) that was achieved in the design and the manufacturing process. It defines the maximum level based only on the designed-in levels of R&M. However, the real-world also has constraints that are not design or manufacturing related (e.g., supply-chain delays that may occur, such as waiting for parts and personnel) and results in the availability that is reflected in real world operations. This availability is known as operational availability, A_o. See Chapter 5 for a quantitative treatment of availability.

Reliability is generally defined as the probability that a system will perform its function(s) as required when used under stated conditions without failure. Estimates of the system's reliability are made early in the design process using a variety of techniques that include using failure data or projections (frequently the area of greatest uncertainty), along with statistical models to make predictions of the system's reliability. These predictions can then be used to assess the design to see if it will meet the system's operational requirements, as well as providing a baseline to evaluate reliability improvements. A system's reliability will be a significant driver of a system's LCC. The measure generally used to measure reliability is the Mean Time Between Failure (MTBF). MTBF is defined as the average time that a device or system is functional before failing. Chapter 3 provides a quantitative discussion of reliability. Maintainability is a measure of the ability to repair or restore a system to an operational condition when maintenance is performed by qualified personnel, using approved procedures and resources. One measure of maintainability is Mean Time to Repair (MTTR). MTTR is defined as the average time from when a device or system fails, the problem is diagnosed, and is repaired so that it is again operational. Chapter 4 provides a quantitative discussion of maintainability. These factors are used to calculate inherent availability.

As the system is developed, there is a balance between its availability and operating and support costs. For a system with a requirement for high availability but with a design that is unreliable, the operating and support cost will be higher, in order to meet the required availability. The support costs increase due to the additional maintenance required, as well as a need to maintain a larger inventory of spare parts. If, on the other hand, sufficient emphasis is placed on R&M during the development, then an appropriate balance can be struck between system readiness and the operating and support costs. Figure 2.6 illustrates this balance.

R&M predictions made during a system's development are not always accurate, and not necessarily static as the system ages. There

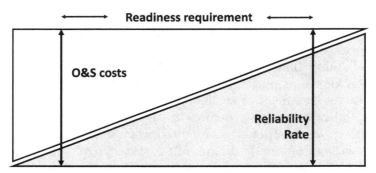

Fig. 2.6. Operating and support costs are corelated to reliability investment (adopted from the GAO).

should be a continuous systems engineering effort to monitor them and look for opportunities to make improvements as the "real world" data accumulates. When the data analysis supports it, additional cost-benefits trade-offs can be made to improve availability, while reducing operating and support costs.

2.4.2 *Choosing the right system architecture – Open versus Closed*

One of the decisions that impacts the design, development, and sustainment of critical systems is the structure of its architecture[2] – open or closed. Open Systems Architecture (OSA) is a system design approach that adopts open standards to produce systems that are inherently interoperable; OSA supports a modular, loosely coupled, and highly cohesive system structure. Said another way, system openness refers to the extent to which system components (e.g., hardware and software) can be independently integrated, removed, or replaced without adverse impact on the existing system. When used appropriately, OSAs provide a degree of flexibility, enabling the integration of rapidly changing technologies, enabling lower-cost modernization through technology insertion. Components can be assembled and connected in a way that the replacement of one component has a minimal effect on the other components in the system. However,

---

[2]System architecture defines how a system is organized, its relationship to all its components, and how they relate to each other and the system's operating environment.

as with all approaches, there are costs as well as benefits. One of the desired outcomes of OSA is to encourage more firms to innovate and develop technology that is interoperable with existing systems.

Closed systems, conversely, consist of unique designs that are tailored for their specific application. As a result, these components can optimize their design to provide the maximum performance with minimum weight, power, and volume. However, their use can result in vendor lock, i.e., making the user dependent on a specific vendor for a proprietary product. Switching to another vendor's product (or software) would impose a substantial, and often prohibitive cost.

In the past, system performance requirements generally compelled that much of the critical functionality in complex electronic systems be developed with program-unique subsystems and architectures. These components required lengthy development schedules and high development budgets. However, in the 1970s, the computer and electronics industries, driven by the competitive pressures of the commercial market and the progression of Moore's Law,[3] began to develop innovative and disruptive technology. The ensuing advancements have enabled the development of increasingly generalizable hardware and software. As a result, once developed, components can be use in a variety of different applications [2.11]. These technological advances have increased the practicality of using OSA [2.12].

Using an OSA, and leveraging existing technologies, often referred to as commercial off-the-shelf (COTS) (discussed in greater detail in the following subsection), can result in cost savings/cost avoidance, facilitate technology refresh, and the ability to rapidly incorporate innovation. At the same time, the proliferation of common component types fosters competition between suppliers. These factors all increase the potential to reduce the systems LCC.

Intuitively, using an OSA appears to be a better strategy, especially from a sustainment and LCC perspective. However, quantifying the actual benefit may be challenging. Program managers will need to perform a business case analysis, examining the engineering trade-offs, to compare and analyze the cost of implementing an open architecture strategy, as opposed to a closed-architecture approach, to determine which is most cost effective.

---

[3]First articulated in 1965, Moore's Law states that the number of transistors on an integrated circuit doubles every 2 years.

U.S. Navy Adopts an OSA to Upgrade Submarine SONAR Systems

The U.S. Navy's Acoustic Rapid COTS Insertion (A-RCI) program implemented an OSA to upgrade its legacy SONAR processing with more capable and flexible commercially available components that were far more capable and flexible than earlier designs. Exploiting this configuration enables a "plug-and-play" capability that allows seamless and efficient upgrades to occur frequently, with little or no impact on submarine scheduling. Using this approach, the Navy could continue to efficiently leverage the advances in the dynamic commercial information technology market, to consistently provide the fleet with near-state-of-the-art processing capability.

The initial A-RCI technology insertions eliminated most of the custom electronic boards used in the system's initial configuration. This immediately and dramatically improved the performance of the operator's displays, increased the system's reliability while keeping development and acquisition costs low. Moreover, programs for the hardware components could now be written in a higher-level language instead of at the tedious assembly level previously required. Programmers could now devote more time to the system's performance, instead of dealing with details of the hardware interface. With the subsequent technology insertions, the program was able to transition to the then-current commercial processors, running at even higher clock speeds. As a result of the increasing reliability of the commercial systems, the Navy introduced a pilot program to test the concept of a Maintenance-Free Operating Period (MFOP) for the A-RCI program. The pilot program's goal was to eliminate the need for maintenance of the A-RCI system while the submarine was underway; all maintenance would be deferred to the next in-port period.

By implementing an OSA, the Navy was able to significantly improve the fleet's SONAR performance by leveraging the rapid advances in commercial computer technology while at the same time keeping development and support costs low. Furthermore, with open systems approach allowed fielded systems to be updated seamlessly and in a cost-effective manner [2.14].

The answer isn't always an OSA; there are reasons to be cautious. Incorporating an OSA will require business and engineering trade-offs, which may reduce the effectiveness of the system, as well as change the incentive structure for the industry partners. The key to the design of open systems is the use of open standards that are well defined, mature, widely used, and readily available. The use of well-defined standards promotes smooth interfacing both within and between systems. If standards do not exist for a new product, it may be best to maintain a closed-system architecture until standards are created [2.13]. Additionally, for legacy systems, redesigning the architecture may be too costly. It may be more cost-effective to keep the existing approach. Finally, if there is only one qualified vendor, developing an open architecture may have cost and performance penalties without any benefits. In summary, there are many reasons to consider an OSA for complex systems, but it should not be the default solution. A detailed cost tradeoff study for open versus closed systems that uses the A-RCI as an example appears in [2.15].

2.4.3 *Commercial off the shelf, modified off the shelf, government off the shelf*

As discussed in Chapter 1, in the post-World War II era, much of the nation's technical innovation was sponsored and funded by the federal government and targeted at national security needs. Now, the commercial marketplace – not the DoD – drives the direction and rate of innovation and development of many technologies that are critical to modern complex systems.

When government is developing critical systems, leveraging these commercial, technical innovations has many potential benefits. These benefits include:

- Using the latest commercially available technology can enable incorporating cutting-edge technology while significantly reducing lead times. As the technology continues to evolve, the system can be updated through subsequent updates.
- The high volume and market competition of commercial off the shelf (COTS) offers not only faster response and friendly user interfaces but lower costs since the development, manufacturing, and support costs are amortized over a much larger customer base.

- When commercial hardware is used, programs can exploit commercially developed (or open-source) operating systems, device drivers, and libraries of applications.
- Using commercially available technologies has the effect of creating a much broader business base (permitting the purchase of systems and equipment from a larger number of vendors). This competitive environment will increase innovation as well as help to ensure continuous price competition.

In summary, greater use of these commercial technologies can shorten development cycle times, reduce LCCs, improve reliability and availability. In addition, by using mature commercial technology, program costs, schedule, and performance requirements can be projected and maintained more easily. In some cases, however, it may be necessary to modify COTS components to meet performance requirements, reducing their benefits. Recognizing these benefits, Congress mandated the requirement to consider and use COTS components for federal government programs [2.16].

U.S. Army Buys COTS Helicopter

The U.S. Army needed to replace aging UH-1 and OH-53C, which were being retired. These helicopters perform a variety of non-combat missions to include logistics and support missions within the U.S. for homeland security, disaster response missions, and medical evacuations. To meet the required schedule, the Army opted for a COTS solution to significantly shorten the acquisition cycle, eliminate development costs, lower acquisition costs, as well as reduce life-cycle logistics and support costs.

The program was limited to procuring aircraft that were already available from commercial sources, and four competitive proposals were submitted. The Army selected the UH-145 proposed by European Aeronautic Defense and Space (EADS) Company North America as its next-generation Light Utility Helicopter (LUH) in late June 2006. Using a COTS solution was a significant paradigm shift for the Army since they generally develop unique military platforms.

(Continued)

(*Continued*)

In the end, the Army demonstrated that these aircraft could meet all the mission requirements, with minimal modifications – and at a fraction of the cost of developing a new helicopter. The first UH-72As were delivered to the Army less than 6 months after contract award with a total RDT&E cost of $3.3M (the UH-60 took over 6 years to develop with an RDT&E cost of $1.698B in 2007 dollars). The limited initial rate of production was one helicopter per month. The production rate increased to two per month and peaked at five per month in 2009. Convinced that the aircraft were reliable and met all requirements, the Army granted EADS the ability to go to full-rate production in August 2007 [2.16], a little more than a year after the initial contract award.

Using COTS products is not without its drawbacks. COTS developers and vendors are driven by today's fast-paced market (characterized by highly volatile business strategies and market positions), not by any specific program's requirements, which may have little, if any, impact. Commercial items and their availability are determined by profit and market share – as a result, designs, and processes, are continuously changing. Moreover, the typical COTS suppliers' product life cycle may be as short as 1–3 years, much shorter than the life of any critical system, which as previously stated may be in service for decades. Consequently, there must be numerous technology refresh and insertion points.

Moreover, vendors may go out of business, merge with other companies, drop products – sometimes without any warning. As a result, change is a constant. This evolving environment can result in inconsistent and short-term availability, obsolescence of components, and unplanned integration and testing requirements. Consequently, programs may require funding for technology updates so the program can insert newer, higher-performing, and often less-expensive components. The rate of change, coupled with many different configuration permutations, requires that programs pay increased attention to configuration management.

COTS software has some unique challenges. With commercially developed software, program staff may lack of insight into the code details and, as a result, may have less understanding of the code than they would have with internally developed software. Additionally, since many commercial software products are large and complex, they are often comprised of millions of lines of source code. This level of complexity precludes a complete, unambiguous analysis of the code for security problems.

In some cases, it may be necessary to modify COTS components to meet performance requirements [2.17, 2.18] or to make the COTS item compatible with the rest of the system (of course these should be kept to a minimum to obtain the maximum benefits). In these cases, the item would be adapted from a COTS item into modified off the shelf (MOTS), generally applied to a software solution that can be modified and customized after being purchased from the software vendor. MOTS is a software delivery concept that enables source code or programmatic customization of a standard prepackaged, market-available software.

Defense Logistics Agency uses a MOTS Solution

The Defense Logistics Agency (DLA) is the department of Defense's combat support agency responsible for the distribution, reutilization, marketing, disposal, tracking, and storing of over five million items of inventory that supply the military services and several civilian agencies with the critical resources they need to accomplish their worldwide missions. In 1999, the DLA tasked itself with completely renovating its IT support system and replacing the DLA's mainframe-based legacy systems – the initiative was named the Business System Modernization (BSM) Program. The BSM program enables the DLA to track product deliveries, budgets, demand projections, and supply schedules in real-time. The DLA made the decision to use a COTS-based solution to leverage commercial development and future upgrades as well as, to the extent possible, to use the commercial best practices embedded in the software.

(Continued)

(Continued)

The new system is based on several COTS systems. The first was an Enterprise Resource Planning (ERP) system developed by SAP. The ERP is responsible for fulfilling orders, procuring the supply items, and managing the finances. DLA integrated another COTS system – Manugistics' Advanced Planning and Scheduling System (APS), to provide a demand and supply planning capability. The processes embedded in these systems were used whenever possible but were modified to accommodate unique DoD Requirements. The new system was successfully deployed, catalyzing dramatic improvement in DLA's performance–the modified COTS technology was a critical element [2.16].

Government off-the-shelf (GOTS) is a term for software and hardware products that are ready to use, and which were developed and are owned by a government agency. Typically, GOTS are developed by the technical staff of the government agency for which it is created. It is sometimes developed by an external entity, but with funding and specification from the agency.

Although the definitions in this section were developed by the federal government, they are now used widely by industry in general [2.19]. While COTS is not the answer to every requirement, with COTS, programs can leverage the massive technology investments of the commercial market and gain the benefits of reduced cycle times, faster insertion of new technologies, lower LCCs, greater reliability and availability, and support from a robust industrial base. These benefits can only be attained, however, by understanding and addressing the new challenges that are created. These challenges are caused by the fundamental difference between designing, developing, and building mission unique components, subsystems, and systems; and integrating COTS solutions. As a result of this difference, the entire acquisition cycle is impacted – from requirements definition through sustainment; all of these must be adjusted to gain the benefits of COTS.

2.5   Technical Data Packages

Technical data packages (TDPs) are critical components of the digital supply chain infrastructure; the TDPs capture and provide critical design information for the development and sustainment of acquisitions, operations, and product support objectives. As supply chains become increasingly optimized with greater transparency across manufacturing and maintenance activities, TDPs will have a greater scope of use in providing the foundational design data for maintenance planning, sustainment strategy development, establishing configurations management requirements, and training for engineering and product support processes [2.20].

In the context of acquisition, technical data is defined as "recorded information, regardless of the form or method of the recording, of a scientific or technical nature" [2.21]. Of particular interest are "form, fit, and function data", defined as "technical data that describes the required overall physical, functional, and performance characteristics of an item, component, or process to the extent necessary to permit identification of physically and functionally interchangeable items" [2.22]. A collection of technical data that fully and authoritatively describes an item in a manner that is "adequate for supporting an acquisition strategy, production, and engineering and logistics support" is referred to as a TDP [2.23].

Access to a system's TDP may be required to support the installation, operation, maintenance, repair, test, and, or modification of a system. However, the procurement of a system does not infer that the procuring organization necessarily has access or rights to the TDP for the system. The acquisition of technical data, including the legal right to use and, or distribute that data, is fraught with many challenges:

- The price of the TDP may be prohibitively high.
- The system developer may have no intention of selling the TDP to the customer at any price.
- The customer's need for the TDP may be in the future and may be uncertain.
- The customer may have purchased TDPs in the past and never used them.

The challenges above demonstrate that it is fundamentally difficult for the customer to determine the value of the TDP. See Section 10.4.1 for discussion of a real-options based TDP acquisition model.

2.6 Production

Product quality,[4] and as a result reliability and sustainability are directly impacted by the reliability of materials used for manufacturing, as well as the reliability of tools, machines and equipment and production processes used. In the past, quality was achieved by conducting a final systemic inspection, which is not very effective. As competition and customer expectations increased over time, manufacturing quality has come to be an absolute requirement. For example, when U.S. made products (automobiles, televisions, and other electronics, etc.) began to lose market share, it was attributed to their poor quality compared to higher quality foreign made products [2.25]. This drove U.S. firms to focus on the quality of their products and to adopt total quality management practices. These practices are focused on the elimination, or prevention, of issues that negatively impact quality and help to ensure the products produced are of high quality. The quality practices permeate virtually all critical design and engineering elements during development, the transition to production, and production itself [2.26]. The objective is incorporate quality into the design of products and processes prior to their manufacture and delivery.

With the development of the Toyota production system (see the inset), Toyota discovered that a smaller number of workers could produce cars using less inventory, less investment, and higher quality (fewer mistakes). This process evolved into what has come to be known as "lean manufacturing", a production process that gives top priority to quality control. The focus on lower inventories enables the

---

[4]The ISO defines quality as "The totality of features and characteristics of a product or service that bear on its ability to satisfy stated or implied needs" [2.24]. Said another way, the quality of a product is a measure of how its inherent characteristics compare to its set of requirements.

identification of problems sooner, putting an increased emphasis on tool and equipment maintenance. Embedded in this concept is the idea of continuous process improvement; most of the improvements originate with the production floor workers. There are other process improvement strategies, such as Six Sigma.[5] Companies that have implemented these concepts have been able to significantly improve product quality and production efficiency [2.28].

Toyota, a Pioneer in Quality Manufacturing

The Toyota Motor Corporation was a pioneer in the quality movement. The company developed the Toyota Production System (TPS), an integrated socio-technical system developed by Toyota that comprises its management philosophy and practices. The TPS system is a significant precursor of the more generic methodology known as lean manufacturing. TPS is steeped in the philosophy of the complete elimination of all waste, embodies all aspects of production in pursuit of the most efficient methods that evolved over many years of trial-and-error efforts to improve efficiency and quality. The process focuses on the elimination of all waste and is based on two pillars:

- Just in Time – Improving productivity by producing "what is needed, when it is needed, and in the amount needed"
- "Jidoka" – the concept that production equipment or operations stop whenever there is an abnormal or defective condition. Production line workers can stop the production line at any time one of them detects a problem.

Based on these simple concepts Toyota Motor Corp. produces high quality vehicles efficiently, that fully satisfy their customers' requirements [2.27].

---

[5]Six Sigma is a set of techniques for process improvement; the goal is to improve the manufacturing process in which 99.99966% of all opportunities to produce some feature of a part are statistically expected to be free of defects. The Six Sigma process improvement strategy attempts to identify and remove the causes of defects and minimizing impact variability in manufacturing and business processes.

Poor manufacturing quality results in rework, scrap; perhaps most importantly from a critical system perspective, product failures. Whatever the strategy used, identification and elimination of failure modes are critical. The Failure Modes and Effects Analyses (FMEA) process is a commonly used structured risk-based methodology for analyzing and preventing failures in manufacturing and assembly processes. The objectives of FMEAs are to identify the potential failure modes and the effects of those failures, rank risk probability and consequence associated with failure modes, and develop actions that will mitigate or eliminate the probability the failure and, or its consequence. Conducting FMEA during the production process development enables the early identification and resolution of problems.

2.7 Best Practices

The acquisition of sustainable systems requires an early focus on requirements, ensuring those attributes that impact availability and operating and support costs are equal in importance to requirements for its performance and acquisition cost. The inclusion of sustainment requirements will require engagement and decisions from all the stakeholders to include policymakers, system developers, and users to ensure their performance needs while minimizing the demand on future resources. Today, critical systems are often designed and developed with the primary focus on performance and cost, without enough consideration of its lifetime sustainment impacts. They must be designed for sustainability.

Long-lived sustainment-dominated systems must have sustainment incorporated into their designs from the beginning of their development. This means that sustainment requirements, that include system availability, reliability, and maintainability, must be considered early enough so that the system design can be impacted (see the Boeing case above). Too frequently, however, sustainment does not receive the requisite attention until it is too late to make the best decisions [2.29].

References

[2.1] GAO (2003). *Setting Requirements Differently Could Reduce Weapon Systems' Total Ownership Costs*, GAO-03-57, February 2003.

[2.2] Moore, J. (2016). Here Are 10 of the Oldest IT Systems in the Federal Government. NextGov. May 25, 2016. Available at: https://www.nextgov.com/cio-briefing/2016/05/10-oldest-it-system s-federal-government/128599/.

[2.3] PMI (2013). *A Guide to the Project Management Body of Knowledge*, 5th Ed. (Project Management Institute Inc.).

[2.4] WMATA (2019). Milestones & History Retrieved from: https://www.wmata.com/about/history.cfm.

[2.5] APTA (2019). Public Transportation Ridership Report Fourth Quarter 2018. American Public Transportation Association. April 12, 2019. Available at: https://www.apta.com/wp-content/uploads/2018-Q4-Ridership-APTA.pdf.

[2.6] Russell, J. (2017). Let's face it: Washington, DC's Metro is the worst in the world. Washington Examiner. June 19, 2017. Available at. https://www.washingtonexaminer.com/lets-face-it-washington-dcs-metro-is-the-worst-in-the-world.

[2.7] Fifield, A. (2016). Nine things about the Tokyo subway that will drive Washington commuters crazy. Washington Post. April 18, 2016. Available at: https://www.washingtonpost.com/news/worldviews/wp/2016/04/18/nine-things-about-the-tokyo-subway-that-will-drive-washington-commuters-crazy/.

[2.8] Repetski, S. (2016). Metro will start running 4000 series cars only in the middle of trains. Might 4000s be completely phased out soon? Greater Washington. November 18, 2016. Available at: https://ggwash.org/view/43562/metro-will-start-running-4000-series-cars-only-in-the-middle-of-trains-might-4000s-be-completely-phased-out-soon.

[2.9] WMATA (2017). 8000 Series Cars DC-12 Design Criteria. January 24, 2017. Available at. https://wmata.com/business/procurement/solicitations/documents/General%20Design%20Criteria%20-%208K%202017%2001%2024.pdf.

[2.10] Wysocki, J., Debouk, R. and Nouri, K. (2004). Shared redundancy as a means of producing reliable mission critical systems. *Annual Symposium Reliability and Maintainability, in Proceedings of RAMS*, pp. 376–381.

[2.11] Guertin, N. H. and Miller, R. W. (1998). A-RCI – The right way to submarine superiority, *Naval Engineers Journal*, 110(2), pp. 21–33.

[2.12] Abbott, J. W., Levine, A., and Vasilakos, J. (2008). Modular/open systems to support ship acquisition strategies", in *Proc. Am. Soc. of Naval Engineers Day* (Arlington, VA).

[2.13] Firesmith, D. (2015). Open System Architectures: When and Where to be Closed. *Software Engineering Institute Blog.* October 19, 2015. Available at: https://insights.sei.cmu.edu/sei_blog/2015/10/open-system-architecture-when-and-where-to-be-closed.html.

[2.14] Gansler, J. S., Lucyshyn, W., and Spiers, A. (2008). Using Spiral Development to Reduce Acquisition Cycle Times. Naval Postgraduate School, September 2008.

[2.15] Chen, S.-P., Sandborn, P. and Lucyshyn, W. (2022). Analysis of the life-cycle cost and capability tradeoffs associated with the procurement and sustainment of open systems, *International Journal of Product Life Cycle Management,* 14(1), pp. 40–69.

[2.16] Gansler, J. S. and Lucyshyn, W. (2008). Commercial-Off-The-Shelf (COTS): Doing It Right, Naval Postgraduate School, September 2008.

[2.17] Wright, M., Humphrey, D., and McCluskey, P. (1997). Uprating electronic components for use outside their temperature specification limits, *IEEE Transactions on Components, Packaging, and Manufacturing Technology, Part A,* 20(2), pp. 252–256.

[2.18] Jensen, F. and Petersen, N. E. (1982). Burn-in: An Engineering Approach to the Design and Analysis of Burn-In Procedures (Wiley, New York).

[2.19] FAR Part 2.101 Definitions. (2020). Effective date: 08-13-2020. Available at https://www.acquisition.gov/content/2101-definitions.

[2.20] Manning, B. (2019). Technical Data Package. AcqNotes. Accessed October 8, 2019. Available at: http://acqnotes.com/acqnote/careerfields/technical-data-package.

[2.21] Department of Defense, Defense Supplement to the Federal Acquisition Regulations (DFARS), Section 252.27-7013, *Rights in Technical Data – Non Commercial Items,* subsection (a)(15).

[2.22] Department of Defense, DFARS, Section 252.227-7013, subsection (a)(11).

[2.23] Department of Defense, Standard Practice, *Technical Data Packages* (MILSTD 31000), Section 3.1.37, November 2009.

[2.24] ANSI/ASQC A8402-1994 (1994). *Quality Management and Quality Assurance – Vocabulary,* American Society for Quality Control.

[2.25] Irland, E. A. (1988). Assuring quality and reliability of complex electronic systems: Hardware and software, *Proceedings of the IEEE,* 76(1), pp. 5–18.

[2.26] GAO (1996). *Commercial Quality Assurance Practices Offer Improvements for DOD,* GAO/NSIAD-96-162, Best Practices. August 1996.

[2.27] Toyota, n.d. Toyota Production System. Retrieved on July 20, 2021, from https://global.toyota/en/company/vision-and-philosophy/production-system/.

[2.28] Klier, T. H. (1993). Lean manufacturing: Understanding a new manufacturing system, *Chicago Fed Letter*, 67.

[2.29] Sandborn, P. and Lucyshyn, W. (2019). Defining sustainment for engineered systems – A technology and systems view. *ASME Journal of Manufacturing Science and Engineering*, 141(2).

Chapter 3

System Failure

A key to predicting and optimizing a system's needed maintenance resources (Chapter 4) is an understanding of the demand for maintenance. The demand for maintenance is primarily driven by the reliability of a system. If one does not understand the reliability of their system, then they are relegated to always performing corrective maintenance, i.e., simply fixing the system after something breaks. The first step to understanding a system's reliability is to understand how and why the system fails.

3.1 How Systems Fail

Failure is defined as the inability of a system to perform its intended function for a specified period of time under specified environmental conditions. Field failures of systems occur for many different reasons. In some cases, there are manufacturing defects that are not detected (or do not become apparent) until the system is in use. There may also be design defects, or system components that simply wear-out as a result of use. Generally, systems fail due to one or more of the following reasons:

- *Wear-out* is deterioration, wear, and/or fatigue over time. For example, car tires, shoes, and carpeting simply wear-out with repeated use. Mechanical systems are prone to wear-out, since moving parts in contact tend to wear and structural elements fatigue. Many electronic systems never reach wear-out; electronic components can wear-out, but in many cases the product is either discarded or fails due to some other cause prior to wear-out occurring. Electronic packaging is more likely to wear-out than the actual semiconductor portions of the system – for example, solder joints can suffer from fatigue cracking with repeated thermal cycling.

- *Overstress* results from unintentionally subjecting a system to environmental stress that is beyond its design specification. An example of overstress would be an electronic system that is struck by lightning.
- *Misuse* is knowingly (intentionally) subjecting a system to environmental stresses that are beyond its design specifications.

Often failures of complex systems are the result of combinations of several problems occurring simultaneously. For example, in 2018–2019 there were several accidents involving the Boeing 737 MAX 8 aircraft. The cause of these crashes was an automated control system. A faulty sensor erroneously reported that the airplane was stalling. The false report triggered an automated system known as the Maneuvering Characteristics Augmentation System, or MCAS. The MCAS system then tried to point the aircraft's nose down so that it could gain enough speed to fly safely, resulting in flying the aircraft into the ground. The failure of the sensor is a problem, but the sensor failure was exacerbated because the 737 MAX 8 pilots did not understand what was happening and therefore, did not or could not override the automated control system. So, this system failed because of a failed sensor compounded by a design (and possibly training) defect.

Note that systems may contain defects or develop defects that are never encountered by their users, either because the users will never use the system under certain environmental stresses or because the function of the system that is impaired is never exercised by the user. In these cases, the defects, although present, never result in system failure and never incur the associated costs or availability penalties of failure or resolution via maintenance.

3.2   Reliability

Reliability is the most important attribute of many types of systems – often more important than cost. Reliability is *quality* measured over time; it is the probability that a system will operate successfully for a specific period of time and under specified conditions when used in the manner and for the purpose intended. High reliability may be necessary in order for one to realize value from the system's performance, functionality, or low cost.

The ramifications of reliability on a system's life cycle are linked to sustainment cost through spare parts requirements and warranty return rates. Indirectly, reliability impacts customer satisfaction, breach of trust, loss of market share, and a host of other factors that influence other costs. The combination of how often a system fails and the efficiency of performing maintenance when a system does fail determine the system's availability (see Chapter 5). The cost of failure avoidance (for example, preventative maintenance) is also linked to reliability.

All of the reliability development and discussion that follows is applicable to the equipment in a factory that produces systems (or the equipment's components), and the final components and systems produced by the factory. However, our primary interest in this book is the reliability of the parts that make up a system, not on the reliability of the factory that makes those parts.

3.2.1 *Failure rate*

If you kept track of all the failures of a particular population of fielded components or systems over its entire lifetime (until every member of the population eventually failed), you could obtain a graph like the one shown in Figure 3.1. Figure 3.1 assumes, that failed system instances are not repaired. We will work exclusively in terms of *time* in this chapter, but in general the *time* axis in Figure 3.1 could be replaced by other usage measures, such as thermal cycles or miles driven.

There are three distinct regions of the graph in Figure 3.1. Early failures due to manufacturing defects (perhaps due to defects induced by shipping and handling, workmanship, process control or contamination) are called infant mortality. The region in the middle of the graph in which the cumulative failures increase slowly is

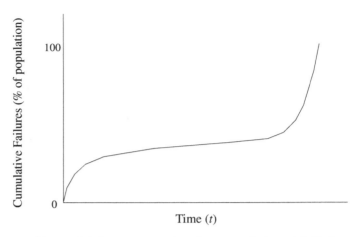

Fig. 3.1.   Observed failures versus time for a population of fielded systems.

considered the useful life of the system. This region is characterized by a nearly constant failure rate. Failures during the useful life are not necessarily due to the way the system was manufactured, but are instead random failures due to overstress and latent defects that don't appear as infant mortality. Finally, the increase in failures on the right side of the graph indicates wear-out of the system due to deterioration (aging or poor or non-existent preventative maintenance). An alternative way to look at the failure characteristics of a system is via the failure rate. Figure 3.2 shows the failure rate that corresponds to the cumulative failures shown in Figure 3.1.[1] Figure 3.2 is known as the "bathtub" curve.

In general, for understanding the maintenance requirements, availability and cost of a system, we care more about a population of systems than we do about the any one particular instance in the population. While the performance of a particular member of the population is interesting, we have to plan, budget, and characterize, based on the whole population.

---

[1]Figure 3.2 is the derivative (slope) of Figure 3.1.

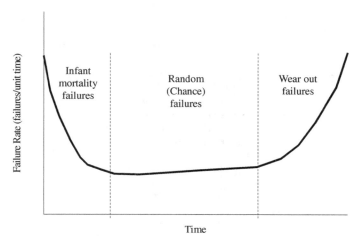

Fig. 3.2. Failure rate versus time observed for a population of fielded systems –
bathtub curve.

3.2.2 *Failure distribution*

To quantitatively describe the failure rate for a population of systems
in terms of reliability, suppose we perform the following test. Start
with 100 instances of a system. All the instances are operational
(unfailed) at time 0. If we subject all the instances to exactly the same
set of environmental stresses, over time the system instances fail, but
they don't all fail at the same time – that is, they are all slightly
different (manufacturing and material variations). This experiment
could give the example data in Table 3.1.

Plotting the fraction of systems failing per time period as a histogram, we obtain Figure 3.3.

3.2.3 *Basic reliability math*

We now wish to add some formal quantitative analysis around the
concept of a failure distribution so as to enable the prediction of
system reliability. If a total of N_0 system instances are tested from
time 0 to time t, the following relation must be true at any time t:

$$N_s(t) + N_f(t) = N_0 \tag{3.1}$$

Table 3.1.   Data collected from environmental testing of $N_0 = 100$ system instances, no repair assumed.

Time period (hours)	Number of systems failing during this time period	Fraction of systems failing during this time period (f)	Total number of systems failed at the end of this time period (N_f)	Total number of systems surviving (unfailed) at the end of this time period (N_s)	Reliability at the end of this time period (R)	Unreliability at the end of this time period (F)	Hazard rate at the end of this time period per 100 hours (h)
0–100	1	0.01	1	99	0.99	0.01	0.010
101–200	3	0.03	4	96	0.96	0.04	0.030
201–300	10	0.1	14	86	0.86	0.14	0.104
301–400	21	0.21	35	65	0.65	0.35	0.244
401–500	31	0.31	66	34	0.34	0.66	0.477
501–600	19	0.19	85	15	0.15	0.85	0.559
601–700	12	0.12	97	3	0.03	0.97	0.800
701–800	2	0.02	99	1	0.01	0.99	0.667
801–900	1	0.01	100	0	0.00	1.00	1.000

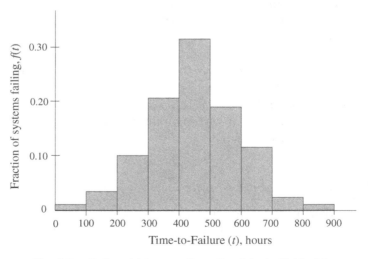

Fig. 3.3.   Failure histogram from the data in Table 3.1.

where $N_s(t)$ is the number of system instances that survived to t without failing and $N_f(t)$ is the number of system instances that failed by t.

If none of the system instances were failed at time 0 ($N_f(0) = 0$), the probability of no failures in the population of system instances from time 0 to time t is given by

$$R(t) = \Pr(T > t) = \frac{N_s(t)}{N_s(0)} = \frac{N_s(t)}{N_0} \qquad (3.2)$$

where T is the failure time. In Equation (3.2), if $N_s(t) = 0$ at some time t, then the probability of no failures at time t is 0. Alternatively, if $N_s(t) = N_0$ at some time t, then the probability of no failures at time t is 1 (100%). Alternatively, the probability of one or more failures between 0 to t is given by

$$F(t) = \Pr(T \le t) = \frac{N_f(t)}{N_0} \qquad (3.3)$$

$R(t)$ is known as the reliability and $F(t)$ is the unreliability of the system at time t. The cumulative failures plotted in Figure 3.1 is $F(t)$. Equations (3.1) through (3.3) imply that for all t,

$$R(t) + F(t) = 1 \qquad (3.4)$$

The reliability $R(t)$ can be constructed graphically from Figure 3.1, as shown in Figure 3.4.

The fraction of failures at time t, $f(t)$, plotted in Figure 3.3, is known as a failure distribution; in general, it is a probability

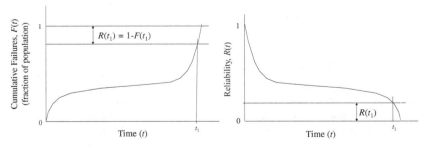

Fig. 3.4. Reliability as a function of time.

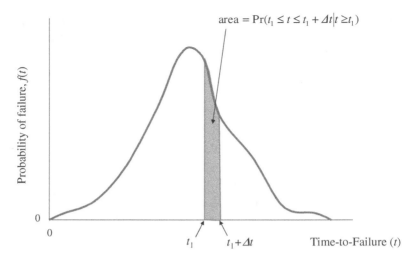

area $= \Pr(t_1 \le t \le t_1 + \Delta t | t \ge t_1)$

Probability of failure, $f(t)$

0

0

t_1 $t_1 + \Delta t$ Time-to-Failure (t)

Fig. 3.5.   Failure distribution.

distribution function (PDF).[2] Figure 3.5 shows what the failure distribution could look like in this case. The area under the probability distribution up to time t_1 (to the left of time t_1) is the probability that the part will fail between 0 and t_1, which is the unreliability $F(t_1)$. Therefore, the area under the $f(t)$ curve to the right of t_1 is the reliability. In general,

$$F(t) = \int_0^t f(\tau)d\tau \qquad (3.5)$$

and therefore, the area under the $f(t)$ curve to the right of t is the reliability, given by

$$R(t) = 1 - F(t) = 1 - \int_0^t f(\tau)d\tau \qquad (3.6)$$

---

[2] Technically you can't convert a histogram (like Figure 3.3) to a PDF because you don't have enough information to do it. However, you can find a distribution that "looks like" the histogram perform a fit to find the parameters, then claim that it is the PDF. For example, you can calculate the mean and standard deviation of a histogram that can be used to determine a normal distribution. If you have the data used to create the histogram there are numerical approaches, e.g., kernel density estimation, which can numerically produce a distribution.

Equation (3.5) is the definition of the cumulative distribution function (CDF). The unreliability is the CDF that corresponds to the probability distribution, $f(t)$. Taking the derivative of Equation (3.6), we obtain

$$\frac{dR(t)}{dt} = -f(t) \tag{3.7}$$

The area within the slice of the distribution between t_1 and $t_1 + \Delta t$ in Figure 3.5 is the probability that a part will fail between t_1 and $t_1 + \Delta t$ when it has already survived to t_1, which is given by

$$\int_{t_1}^{t_1+\Delta t} f(\tau)d\tau = F(t_1 + \Delta t) - F(t_1) = R(t_1) - R(t_1 + \Delta t) \tag{3.8}$$

The failure rate is defined as the probability that a failure occurs in the time interval, given that no failure has occurred prior to the start of the time interval:

$$\frac{R(t) - R(t + \Delta t)}{\Delta t R(t)} \tag{3.9}$$

Taking the limit as Δt goes to 0 and using Equation (3.7), Equation (3.9) gives the hazard rate, or instantaneous failure rate:

$$h(t) = \lim_{\Delta t \to 0} \frac{R(t) - R(t + \Delta t)}{\Delta t R(t)} = -\frac{1}{R(t)}\frac{dR(t)}{dt} = \frac{f(t)}{R(t)} \tag{3.10}$$

The hazard rate is a conditional probability of failure in the interval t to $t + dt$, given that there was no failure up to time t. Restated, hazard rate is the number of failures per unit time per the number of non-failed systems surviving at time t. Figure 3.2 is a plot of the hazard rate. Using Equation (3.10), the reliability is given by,

$$R(t) = e^{-\int_0^t h(\tau)d\tau} \tag{3.11}$$

Once a system has past the infant mortality (or early failure) portion of its life, it enters a period during which the failures are random due to changes in the applied load, overstressing conditions, and variations in the materials and manufacturing of the system.[3]

---

[3]Burn-in is used to accelerate early failures so that systems are beyond the infant mortality portion of the bathtub curve before they are shipped to customers (see Section 3.2.7).

Depending on the type of system or component, different distributions can be used to model the reliability during the random failure (field use) portion of the system's life. The following sections describe two commonly used failure distributions that are used to describe the failure rates of systems.[4]

3.2.4 *Constant failure rate (exponentially distributed time-to-failure)*

The simplest assumption about the field-use (random failures) portion of the life of a system is that the failure rate is constant:

$$h(t) = \lambda \tag{3.12}$$

Using Equations (3.10) and (3.7), we can solve for the PDF in this case:

$$f(t) = h(t)R(t) = \lambda - \lambda \int_0^t f(\tau)d\tau \tag{3.13}$$

Taking the derivative of both sides of Equation (3.13) gives us

$$\frac{df(t)}{dt} = -\lambda f(t) \tag{3.14}$$

Equation (3.14) is satisfied by

$$f(t) = \lambda e^{-\lambda t} \tag{3.15}$$

where $f(t)$ is an exponential distribution. The corresponding CDF (unreliability) and reliability are given by

$$F(t) = \int_0^t \lambda e^{-\lambda \tau}d\tau = 1 - e^{-\lambda t} \tag{3.16}$$

$$R(t) = 1 - F(t) = e^{-\lambda t} \tag{3.17}$$

The mean of $f(t)$ is given by the expectation value of $f(t)$:

$$\mathrm{E}[T] = \int_0^\infty tf(t)dt = \int_0^\infty t\lambda e^{-\lambda t}dt = \frac{1}{\lambda} \tag{3.18}$$

---

[4]Many other distributions can be used. Readers can consult any reliability engineering text for information on other distributions.

$E[T]$ is also known as the mean time to failure ($MTTF$) or, if the failed systems are repaired to "good as new" condition instantaneously after each failure, the $E[T]$ is the mean time between failures ($MTBF$). Note that at $t = MTBF = 1/\lambda$, $R(t) = 1/e = 0.37$. This means that $F(t) = 1 - 0.37 = 0.63$ or 63% of the population has failed by $t = MTBF$.

A Failure In Time (FIT) rate, which is quoted for many systems, is a constant failure rate.[5] The exponential distribution assumes that systems fail at a constant rate, regardless of accumulated age.[6] This is not a good assumption for many real applications and $MTBF$ has been used for some systems as a reliability specification without realizing that in most cases it is almost impossible to demonstrate. The assumption of a constant failure rate ignores the effect of system design, system assembly and manufacturing procedures, component interaction, overstress and temperature cycling. Describing a system using only an $MTBF$ (which is often done) is problematic. An $MTBF$ can be derived if the distribution is known, but the distribution cannot be derived, calculated, or inferred from only a mean value, and calculating the probability of any other event occurring is impossible without knowing the underlying distribution.[7]

3.2.5 *Weibull time-to-failure distribution*

The Weibull distribution is a more widely used to describe the failure rates in reliability engineering than exponential distributions because of the flexibility it has in accommodating different forms of the

---

[5] A FIT rate is failures per billion (10^9) hours.

[6] The exponential distribution is often referred to as memoryless because the past has no bearing on its future behavior. Every moment in time is essentially a good-as-new restart for the system or component regardless of how much time has passed.

[7] In some cases, the use of an exponential distribution may indicate the use of a reliability prediction model that is not based on actual data, but rather utilizes compiled tables of generic failure rates (exponential failure rates) and multiplication factors (e.g., for electronics, MIL-HDBK-217 [3.1]). These analyses provide little insight into the actual reliability of the systems in the field [3.2].

hazard rate. The PDF for a three-parameter Weibull is given by

$$f(t) = \frac{\beta}{\eta} \left(\frac{t - \gamma}{\eta} \right)^{\beta - 1} e^{-\left(\frac{t-\gamma}{\eta} \right)^{\beta}} \tag{3.19}$$

where β is the shape parameter, η is the scale parameter, and γ is the location parameter (also sometimes referred to as the failure free operating period, see Section 4.10). Note, Equation (3.19) is only valid for $t \geq \gamma$. The corresponding CDF, reliability, and hazard rate are given by

$$F(t) = 1 - e^{-\left(\frac{t-\gamma}{\eta} \right)^{\beta}} \tag{3.20}$$

$$R(t) = e^{-\left(\frac{t-\gamma}{\eta} \right)^{\beta}} \tag{3.21}$$

$$h(t) = \frac{\beta}{\eta} \left(\frac{t - \gamma}{\eta} \right)^{\beta - 1} \tag{3.22}$$

With an appropriate choice of parameter values, the Weibull distribution can be used to approximate many other distributions. For example, $\beta < 1$ represents a decreasing failure rate (infant mortality), $\beta = 1$ constant failure rate (exponential distribution) and $\beta > 1$ increasing failure rate (wear-out). $\beta = 3$ approximates a normal distribution and $\beta = 2$ gives a Rayleigh distribution.

3.2.6 *Conditional (mission) reliability*

Conditional reliability is the probability that a system will survive for an additional time t given that it has already survived up to time T. The system's conditional reliability function is given by

$$R(t, T) = \frac{R(t + T)}{R(T)} \tag{3.23}$$

$R(t, T)$ is also called a "mission reliability" because it can be used to represent the probability of completing a mission successfully given that the system is not good-as-new. For example, if $R(20) = 0.4$ and $R(10) = 0.6$ then $R(10, 10)$, the probability of survival for an additional 10 time units given that the system has already survived 10 time units is 0.67.

For a constant failure rate, the $R(t, T)$ is,

$$R(t, T) = \frac{R(t + T)}{R(T)} = \frac{e^{-\lambda(t+T)}}{e^{-\lambda T}} = e^{-\lambda t} = R(t)$$

which makes sense because a constant failure rate implies that the system is always good-as-new no matter how much time has passed.

3.2.7 *Stress screening*

There are numerous approaches by which systems can be stressed prior to being placed into service (and often, prior to completing the manufacturing process). The goal of this is to identify particular units that would fail during the initial, high-failure rate infant mortality phase of the bathtub curve shown in Figure 3.2. The objective is to make the stress-screening period sufficiently long (or stressful) that the unit can be assumed to be mostly free of early failure risks before it is placed into service. Screening approaches can be broadly classified as follows:

- Burn-in is generally a lengthy process of *powering* a system at a specified *constant* stress level within the system's specification limits. The goal of burn-in is to expose the infant mortality of systems.
- Environmental Stress Screening (ESS), which evolved from burn-in, is a process of stressing a system in continuous cycles between predetermined environmental extremes. ESS is primarily temperature cycling plus random vibration. ESS differs from burn-in because it exposes systems to environmental stresses outside of the system's specification limits. While ESS may be used to expose infant mortality, it may also target forcing latent defects (that are not part of infant mortality) to manifest themselves. Highly Accelerated Stress Screening (HASS) is similar to ESS but may include stresses that are not experienced in the field to accelerate relevant defects.
- Other screening methods. Discriminator screening uses a discriminator comprised one or more parameters of the product to identify

systems with intrinsic defects. Degradation screening subjects systems to elevated stress levels for prescribed lengths of time and assesses the degradation rate of the systems.

In this section, we are concerned with discovering infant mortality, which we will refer to generally as screening. A precondition for successful screening is a bathtub-curve failure rate, meaning that there are a non-negligible number of early failures (infant mortality), after which the failure rate decreases. Stressing all units for a specified time causes the units with the highest failure rate to fail first so they can be taken out of the population. The units that survive the screen will have a lower failure rate thereafter.

The strategy behind burn-in type screening (see Figure 3.6) is that infant mortality system failures can be avoided at the expense of performing the screen and a reduction in the number of units shipped to customers.[8]

Screening (or the amount of screening done) is a cost tradeoff. Performing the screen costs money that must be recouped through lower warranty returns or maintenance costs. Several models exist that can be used to estimate the optimal amount of screening to do, i.e., the amount of screening that minimizes the life-cycle cost of the system. Consider the following model from Ebeling [3.3]. The expected number of total failures during screening (n_u total units start the screen) is given by,

$$n_u[1 - R(t_{bd})] \tag{3.24}$$

The conditional reliability that a unit will last through a period of time T_W (T_W could represent the warranty period for the system,

---

[8]The view of burn-in in the electronics industry has changed significantly in the past 20–30 years. Twenty years ago, burn-in was an important process in the electronics industry due to high infant mortality rates. Back then, you had to make a case NOT to include a burn-in in your process. These days the opposite is true – in many industries the case must be made for burn-in due to the cost implications and reasonably low infant mortality rates.

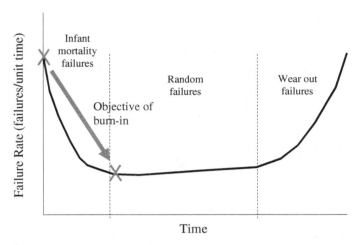

Fig. 3.6. The goal of burn-in type screening is to reach the random failures portion of the bathtub curve before sending the system to the customers.

see Section 4.4.1.2) given that it survives the screen (t_{bd}) is[9]

$$R(T_W|t_{bd}) = \frac{R(t_{bd} + T_W)}{R(t_{bd})} \tag{3.25}$$

So, the probability of failure during T_W is,

$$1 - \frac{R(t_{bd} + T_W)}{R(t_{bd})} \tag{3.26}$$

and the expected number of total failures during T_W is[10]

$$n_u R(t_{bd}) \left(1 - \frac{R(t_{bd} + T_W)}{R(t_{bd})}\right) = n_u(R(t_{bd}) - R(t_{bd} + T_W)) \tag{3.27}$$

where $n_u R(t_{bd})$ is the number that survived the screen. In this case, the expected total cost of the will be

$$C_{\text{total}} = n_u[C_{Bt}t_{bd} + C_P(1 - R(t_{bd})) + C_{cw}(R(t_{bd}) - R(t_{bd} + T_W))] \tag{3.28}$$

---

[9]This model also assumes that there is no acceleration of the stress conditions in the screening process, i.e., this would technically be referred to as a "burn-in".
[10]Equation (3.27) represents only first failures. This could be re-stated using a renewal function (see Section 4.4.1.2) to represent all failures in T_W.

where C_P is the unit cost, C_{cw} is the cost of resolving a warranty claim, and C_{Bt} is the screening cost. The first term in the cost to perform the screen; the second term is the cost of units lost during screening; and the third term is the cost of field failures for units that survived screen. The length of the screen can be optimized by setting the derivative of Equation (3.28) with respect to t_{bd} equal to zero,

$$\frac{dC_{\text{total}}}{dt_{bd}} = n_u \left[C_{Bt} - C_P \frac{dR(t_{bd})}{dt_{bd}} \right.$$

$$\left. + C_{cw} \left(\frac{dR(t_{bd})}{dt_{bd}} - \frac{dR(t_{bd} + T_W)}{dt_{bd}} \right) \right] = 0 \quad (3.29)$$

If, for example, the reliability is given by a Weibull in Equation (3.21) with $\gamma = 0$, the derivative of the reliability is given by,

$$\frac{dR(t_{bd})}{dt_{bd}} = \frac{d}{dt_{bd}} e^{-\left(\frac{t_{bd}}{\eta}\right)^{\beta}} = -e^{-\left(\frac{t_{bd}}{\eta}\right)^{\beta}} \frac{d}{dt_{bd}} \left(\frac{t_{bd}}{\eta}\right)^{\beta}$$

$$= -e^{-\left(\frac{t_{bd}}{\eta}\right)^{\beta}} \frac{\beta t_{bd}^{\beta-1}}{\eta^{\beta}} \quad (3.30)$$

As an example, consider a specialized fuel pump that has a high infant mortality rate [3.4]. The manufacturer provides a 25-operating-hour warranty. The screening costs $50/unit/hour to run (C_{Bt}); the cost of a unit is $350 (C_P) and; the cost of a warranty claim is $1200/claim (C_{cw}).[11] Assume that we are only dealing with first failures (this is a reasonable approximation if the warranty period is short). Assuming the reliability of the pump is given by a 2-parameter Weibull distribution with $\beta = 0.30098$ and $\eta = 472.3$ operating hours $(\gamma = 0)$. What is the optimal (minimum cost) screening period (t_{bd})? Solving Equation (3.29) using Equation (3.30) gives a burn-in of $t_{bd} = 0.54366$ operational hours, this value of t_{bd} minimizes the life-cycle cost of this system.

Many other models for burn-in optimization and costing exist. Nguyen and Murthy [3.5] and Chien and Sheu [3.6] determined

---

[11] Assuming that the fixed/non-recurring cost of burn-in is 0.

the optimal burn-in time by minimizing the expected cost for repairable and non-repairable systems sold under various warranty policies. Kim and Kuo [3.7] studied the systems that were repaired during the burn-in procedure and put back into the test chamber to continue undergoing the burn-in process. They developed a probabilistic model, which was useful for a two-level[12] burn-in procedure to optimize reliability and the economy of production when compatibility existed in components, as well as in connection. Sandborn [3.8] provides a similar cost model for screening that includes the calculation of a return on investment for the screening process.

3.2.8 *Multiple failure mechanisms*

There are usually multiple different mechanisms that can cause a component to fail. If $f_1(t)$ and $f_2(t)$ represent the time-to-failure distributions associated with mechanisms 1 and 2 (and the failure mechanisms are independent), then the time-to-failure probability density function for the component is given by

$$f(t) = f_1(t)[1 - F_2(t)] + f_2(t)[1 - F_1(t)] \qquad (3.31)$$

where the $F(t)$ are given by Equation (3.5).

3.2.9 *The reliability of systems*

The reliability of a system depends on the reliability of the system's individual components and how those components are organized. Two fundamental arrangements are possible: series and parallel (Figure 3.7).

Series systems are systems that only operate satisfactorily if all the components in the system are operating. In a series system, the failure of any component causes the whole system to fail. Since reliability is a probability, in a series connection, the system reliability

---

[12]A two-level burn-in process refers to screening at both the component and module levels. This approach allows screening for assembly defects.

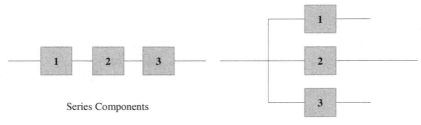

Series Components

Parallel Components

Fig. 3.7.   Series and parallel components.

is simply the product of the individual component reliabilities,

$$R(t) = R_1(t)R_2(t) \cdots R_n(t) \tag{3.32}$$

If the components are connected in parallel, then the system operates satisfactorily if one or more of the parallel components in the system are operating. Parallel components represent "redundancy" in the system. In engineering, redundancy is defined as the duplication of critical components or functions of a system with the intention of increasing the reliability of the system. In this case the system unreliability is

$$F(t) = [1 - R_1(t)][1 - R_2(t)] \cdots [1 - R_n(t)] \tag{3.33}$$

From Equation (3.4), the system reliability is

$$R(t) = 1 - F(t) = 1 - [1 - R_1(t)][1 - R_2(t)] \cdots [1 - R_n(t)] \tag{3.34}$$

Other system configurations that are composed of both series and parallel components are aggregates of Equations (3.32) and (3.34).

3.2.10   *The cost of reliability*

As highlighted in Chapter 2, reliability isn't free. The cost of providing reliable systems includes the costs associated with designing and producing a reliable system, testing the system to demonstrate the reliability it has, and creating and maintaining a reliability organization. The more reliable the system is, the less money will have to be spent after manufacturing on sustaining the system. Reliability is,

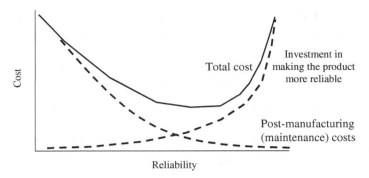

Fig. 3.8. The relationship between reliability and cost.

however, a tradeoff and there may be an optimum amount of effort that should be expended on making systems reliable, as shown in Figure 3.8.

The post-manufacturing (maintenance) costs in Figure 3.8 can represent different things depending on the type of system and the portion of its life cycle that it is in. For a consumer product, this often manifests itself as a warranty cost, i.e., the more the reliable the product, the smaller the warranty costs. For other types of systems this is more generally maintenance costs.

3.3 Software Reliability

The concept of software reliability similar to that of hardware reliability. The definition of software reliability is: "the probability of failure-free software operation for a specified period of time in a specified environment". There is no question that software reliability contributes to system reliability.

3.3.1 *How software fails*

A fundamental difference between software and hardware is that software doesn't age or wear out and it isn't subject to failure mechanisms that are driven by physical environment stresses caused by temperature, vibration, humidity, salt, etc. Software usually stays in the same condition as when it was created and unless there are changes caused by hardware – like changes in the storage content or data path – it is not unreasonable to assume that software doesn't actually "break".

Just because software doesn't break, does not mean that it can't fail. Software failure mechanisms include [3.9]:

- Errors in requirements, logic, coding, or documentation
- Ambiguities
- Oversights or misinterpretation of the specification that the software is supposed to satisfy
- Carelessness or incompetence in writing code
- Inadequate testing
- Incorrect or unexpected usage of the software or other unforeseen problems

Failures can also be characterized based on their seriousness,

- Transient versus permanent
- Recoverable versus non-recoverable
- Corrupting versus non-corrupting

Unlike hardware software does not have energy related wear-out, software reliability is not a function of operational time and redundancy cannot improve software reliability if identical software components are used.

3.3.2 *Software reliability*

The bathtub curve for hardware (Figure 3.2) is not applicable for software. In the case of software, the failure intensity follows a relation like the one shown in Figure 3.9.[13] The upgrades in Figure 3.9 are feature upgrades, not upgrades to improve reliability. Generally, when feature upgrades are done, the functionality of the software is increased, thereby increasing its complexity. Drops in the

---

[13]What is an appropriate time unit in Figure 3.9? Depending on the type of software this could be raw execution time (for continuously operated real-time systems), the number of transactions (for transaction-based systems), or a software release count.

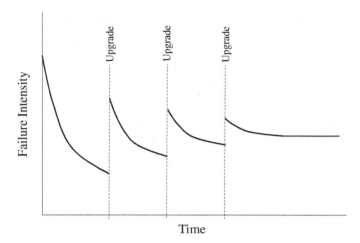

Fig. 3.9.   Failure intensity versus time for software.

failure intensity initially and after upgrades are due to "bug" fixes. Reliability upgrades are possible. These upgrades involve redesigning or reimplementation of some modules using better engineering approaches.

While many models have been proposed (since the 1970s), the problem of quantifying software reliability remains an open question. Several approaches exist:

- *System metrics*: Software size is generally thought to be an indication of complexity, development effort and reliability. Source Lines Of Code (SLOC), or function/feature points can be used to measure software size. Standardized counting, comparison of software written in different languages, code reuse and automated code generation techniques complicate these approaches.
- *System management and process metrics*: Higher reliability can be achieved by using better development processes, risk management processes, configuration management processes, etc.
- *Fault and failure metrics*: A combination of the number of faults found during testing and the failures and other problems reported by users. The data collected is used to calculate failure density,

Mean Time Between Failures ($MTBF$) or other parameters to measure or predict software reliability.

The basic reliability metrics used for software are:

- $MTTF$ = Mean Time To Failure. The time interval between successive failures. This is relevant for systems with long transactions, i.e., CAD systems, word-processor systems.
- $MTBF$ = Mean Time Between Failure. $MTBF$ is given by

$$MTBF = MTTF + MTTR \qquad (3.35)$$

  where $MTTR$ is the Mean Time To Repair, the time to diagnose and repair the problem. In this case the time measurements are in real time (not execution time).
- $ROCOF$ = Rate Of Occurrence Of Failure, the failure intensity. The number of failures occurring in a unit time interval.
- $POFOD$ = Probability Of Failure On Demand. The probability that the system will fail upon a service request. This is an important measure for safety-critical systems where services are demanded occasionally. $POFOD$ is similar to instantaneous availability, Section 5.1.1.
- Availability = Availability is the probability that the system is available for use at a given time (see Chapter 5).

Metrics are easy to define, but many practical difficulties exist. For software, reliability depends not only on the existence of errors, but also on the probability that an error will impact the output or function of the software. The following provides an overview of basic statistical methods for assessing software reliability, see [3.10] for a more detailed treatment.

We assume that errors exist in the software code that contribute to its failure. If the number of errors appearing in time t is $N(t)$ ($N(0) = 0$, and all the errors are independent), then the probability that $N(t)$ equals exactly n errors in time t is given by[14]

$$Pr(N(t) = n) = \frac{[M(t)]^n}{n!} e^{-M(t)} \qquad (3.36)$$

---

[14]Equation (3.36) is a single term from Equation (4.12) that will be developed in the sparing analysis discussion.

where $M(t)$ is the mean number of errors appearing in the interval $(0, t]$. $M(t)$ is generally known as a renewal function, it counts the number of times in an interval $(0, t]$ that a system that is instantaneously repaired fails.[15] See Section 4.4.1.2 for a detailed discussion of renewal functions. For software reliability, many different functional forms of $M(t)$ have been proposed (see [3.10] for an overview). One common form called the Goel-Okumoto (G-O) model [3.11] is

$$M(t) = a_s[1 - e^{-bt}] \qquad (3.37)$$

where a_s is the total number of errors at $t = 0$ and b is a constant (the failure intensity[16]). If errors are fixed as they appear and the fixes do not introduce any new errors, then the number of errors that remain at time t is,

$$N(t) = a_s - M(t) = a_s e^{-bt} \qquad (3.38)$$

The reliability is given by

$$R(t) = e^{-M(t)} \qquad (3.39)$$

The number of renewals in the interval $[t_0, t_0 + t]$ is given by

$$M(t_0 + t) - M(t_0) = a_s(e^{-bt_0} - e^{-b(t_0 + t)}) \qquad (3.40)$$

and the reliability in the interval $[t_0, t_0 + t]$ from Equation (3.23) is

$$R(t, t_0) = e^{-a_s(e^{-bt_0} - e^{-b(t_0 + t)})} \qquad (3.41)$$

---

[15] For a non-repairable system where each system instance is repaired at most one time, $M(t)$ equals $F(t)$ the unreliability.

[16] The terms "failure intensity" and "failure rate" are similar and sometimes used interchangeably, however, the first failure is governed by a distribution $f(t)$ with failure rate $\lambda(t)$, while each subsequent failure is governed by the intensity function of the process. If the system is non-repairable or there is a constant failure rate, then failure rate and failure intensity are the same. In general, the relationship between the failure intensity and the renewal function is given by, $m(t) = (dM(t))/dt$ (where m is also called the renewal density, or denoted by b, the failure intensity, for software systems). The probability that a system of age t fails between t and $t + \Delta t$ is given by the intensity function $m(t)\Delta t$, which is not conditioned on not having any system failures up to time t. Alternatively, the failure rate is conditioned on the system having no failures up to t. So, why does Figure 3.9 use failure intensity – because software is repairable.

One commonly used parameterization of the renewal function is the Musa model [3.10]

$$M(t) = N_0 \left[1 - e^{-\frac{\omega t}{N_0 T_0}} \right] \tag{3.42}$$

where N_0 is the number of errors at $t = 0$, T_0 is the MTTF, ω is the testing compression factor equal to the ratio of the equivalent operating time to the testing time, and t is measured in program execution time.

Consider the following example. A program will experience 100 errors in infinite time. Its initial failure intensity was 10 failures/CPU hour. What is the program's failure intensity at 10 CPU hours? From Equation (3.42) and the definition of $m(t)$ in footnote 16,

$$m(t) = \frac{\omega}{T_0} e^{\frac{-\omega t}{N_0 T_0}} \tag{3.43}$$

At $t = 0$, Equation (3.43) gives, $10 = \omega/T_0$, using $N_0 = 100$, find m at $t = 10$ from Equation (3.43); $m(10) = 3.679$ failures/CPU hour.

As a second example, consider a program that is assumed to contain 300 errors initially (at $t = 0$). The *MTTF* when testing begins is 1.5 hours. The testing compression factor is 4. Test/fix is done to reduce the number of errors to 10. What will the reliability be over 50 hours of running after the testing is done? In this case, $N_0 = 300$, $T_0 = 1.5$ hours, $\omega = 4$, and $M(t) = 300 - 10$, so solving Equation (3.42) for t gives $t = 382.63$ hours. From Equation (3.41), the failure intensity at $t_0 = 382.63$ hours is $b = 0.08889$ failures/hour and the reliability over an additional $t = 50$ hours is 0.0276.

3.4   Quality, Durability, Robustness and Safety

Quality and reliability are related but not the same. Quality is a measure of how well a system performs its function (fit for use, or fitness for purpose), while reliability is a measure of how well a system maintains its quality over time when used under appropriate conditions. The clearest difference is that quality does not depend on time and reliability does. Quality is a static photograph taken at the end of manufacturing and reliability is a movie of the

product over time. Defects in a product at the end of the manufacturing process that escaped detection can negatively affect a system's quality. Defects that develop into problems that negatively
affect the system's operation over time are considered reliability
issues.

Quality is commonly defined as "conformance to requirements
at the start of use". Assuming the system specifications adequately
capture customer requirements, the quality level can be measured by
the fraction of units shipped that meet specifications. The question
is, how many of the units shipped will still meet specifications after
a specified period of time?

Durability is closely related to quality and reliability. Durability is a measure of how long a system will last before it becomes
non-functional (or it becomes impractical to continue repairing it).
Durability is often confused with reliability. The difference is that
reliability is a probability of surviving over some period of time.
Durability allows for maintenance (preventative or corrective) to be
performed on the system and is a measure of how long before the
occurrence of wear-out failures. Durability is the fitness for service
over a suitable period.[17] For example, the majority electronic parts
and systems never reach wear-out, therefore, durability is not an
issue. But this does not mean that electronic parts have 100% reliability. In repairable systems, durability is a measure of the system's
ability to function after repairs, or conversely, the maximum amount
of repair that can be done on the system before retiring it. For a
complex system, durability (as a time) could be measured as

$$\text{Durability} = (MTBF)N_{\max} \qquad (3.44)$$

where $MTBF$ is the mean time between failure and $N_{\max}$ is the
maximum number of repairs of the system (or component) allowed
before the system is retired (or replaced). For some systems, $N_{\max}$
is a pre-determined policy, in other cases, the point where a system
is retired is determined by the cost of repair, i.e., when the cost

---

[17]There are different types of durability. With respect to critical systems (hardware and software) we are generally equating durability to aging. However, for
other products durability may be a toughness measure (e.g., concrete), rot proofing, rust proofing, etc.

to repair the failed system exceeds some threshold fraction of the cost of a new replacement, the replacement is done instead of the repair.

The term "robustness" (sometimes used interchangeably with durability by engineers) implies that a system is designed to be resistant to failure despite partial damage. Reliability is the probability of no failure, robustness is a property of the system's design/architecture. In the context of software reliability, robustness means the degree to which a system is able to continue functioning in the presence of invalid inputs or error other conditions without serious effect, i.e., without crashing or becoming locked in a loop. For hardware, robustness means "the insensitivity of products or processes against different sources of variation, such as production or assembly tolerances, not (fully) specified load scenarios or ambient use conditions" [3.12].

Reliability is also related to safety. Safety can be defined as "freedom from those conditions that can cause death, injury, occupational illness, or damage to or loss of equipment or property, or damage to the environment" [3.13]. Safety is not the same as reliability. Reliability is associated with the probability of failure; safety is associated with the probability of a failure resulting in a bad outcome. Highly reliable systems are often assumed to also be safe; however, reliability does not necessarily infer safety or vice versa. The safest car may be the car that is always broken down and never leaves your driveway – a car that we would view as having poor reliability. Safety is a topic that many reliability engineers steer clear of because safety has legal ramifications, so in their minds, safety does not enter into reliability discussions.

References

[3.1] U.S. Department of Defense (1991). Military Handbook: Reliability Prediction of Electronic Equipment, MIL-HDBK-217F(2).

[3.2] ReliaSoft (2001). Limitations of the exponential distribution for reliability analysis, *Reliability Edge*, 2(3). Available at: http://www.reliasoft.com/newsletter/4q2001/exponential.htm

[3.3] Ebeling, C. E. (1997). *An Introduction to Reliability and Maintainability Engineering* (McGraw-Hill Companies, Inc., USA), pp. 312–315.

[3.4] Soni, S. and Badiru, A. B. (2014). AFIT engineering systems reliability tool, in *Handbook of Industrial and Systems Engineering*, 2nd Ed., A. B. Badiru (ed.) (Taylor & Francis Group, USA).

[3.5] Nguyen, D. G. and Murthy, D. N. P. (1982). Optimal burn-in time to minimize cost for products sold under warranty, *IIE Transactions*, 14(3), pp. 167–174.

[3.6] Chien, Y. H. and Sheu, S. H. (2005). Optimal burn-in time to minimize the cost for general repairable products sold under warranty, *European Journal Operational Research*, 163, pp. 445–461.

[3.7] Kim, W. and Kuo, Y. (2004). Two-level burn-in for reliability and economy in repairable series systems having incompatibility, *International Journal of Reliability*, 11, pp. 197–211.

[3.8] Sandborn, P. (2017). *Cost Analysis of Electronic Systems*, 2nd Ed. (World Scientific, Singapore).

[3.9] Keiller, P. A. and Miller, D. R. (1991). On the use and the performance of software reliability growth models, *Software Reliability and Safety*, 32(1–2), pp. 95–117.

[3.10] Musa, J., Iannino, A. and Okumoto, K. (1987). *Software Reliability Prediction and Measurement* (McGraw-Hill, USA).

[3.11] Goel, A. L. and Okumoto, K. (1979). A time dependent error detection model for software reliability and other performance measures, *IEEE Trans. Reliability*, R-28(3), pp. 206–211.

[3.12] Kemmler, S., Eifler, T., Bertsche, B., and Howard, T. J. (2015). Robust reliability or reliable robustness? – Integrated consideration of robustness and reliability aspects, *Proc. of the 27. Fachtagung Technische Zuverlässigkeit*, pp. 295–306.

[3.13] U.S. Department of Defense (1993). *Military Standard: System Safety Program Requirements*, MIL-Std-882C.

Problems

3.1 Your company manufactures a GPS chip for use in marine applications. Through extensive environmental testing, you found that 5% of the chips failed during a 400-hour test. Assuming a constant failure rate, answer the following questions:

    (a) What is the probability of one of your GPS chips lasting at least 5000 hours?

    (b) What is the mean life (*MTBF*) of the GPS chips?

3.2 The *MTBF* (Mean Time Between Failure) for a voting machine has to be at least 163 hours.

(a) Assuming the time-to-failure is exponentially distributed, what constant failure rate corresponds to an *MTBF* of 163 hours?

(b) If the *MTBF* is 163 hours, what is the probability of a voting machine making it through a 15-hour election day without failing?

(c) If the *MTBF* is 163 hours, what % of voting machines will fail at least once in a 15-hour voting day?

3.3 The time-to-failure of a population of electronic parts is believed to follow a Weibull distribution with $\beta = 0.2$ (shape parameter), $\eta = 200$ hours (scale parameter), and $\gamma = 0$ (location parameter). What is the probability that one of these parts will function for more than 10 years?

3.4 Given a 3-parameter Weibull with $\beta = 2.3$ (shape parameter), $\eta = 72{,}000$ miles (scale parameter), and $\gamma = 10{,}000$ miles (location parameter), create a plot of the distribution of the miles-to-failure, using a spreadsheet.

3.5 Over the last several years your company has collected the following monthly repair frequency history (top of next page) on a key piece of manufacturing equipment. If the average repair cost (parts and labor) is \$356/repair, calculate the expected annual repair budget for this piece of equipment.

Repairs/month	Probability
0	0.02
1	0.05
2	0.08
3	0.1
4	0.12
5	0.13
6	0.12
7	0.16
8	0.11
9	0.09
10	0.02

3.6 Two failure mechanisms are relevant for a component. The first failure mechanism has a constant failure rate of $\lambda = 0.002$ hours^{-1}. The second failure mechanism has a constant failure rate of $\lambda = 0.003$ hours^{-1}. What is the probability that the component will survive at least 600 hours of operation? Assume that the two failure mechanisms are independent.

3.7 A system consists of five components with the following *MTBF*s:

Component A: $MTBF = 11{,}540$ hours
Component B: $MTBF = 16{,}400$ hours
Component C: $MTBF = 7500$ hours
Component D: $MTBF = 11{,}800$ hours
Component E: $MTBF = 6600$ hours

The five components are connected in series (i.e., all have to work for the system to work).

(a) Assuming constant failure rates for all the components, what is the reliability of the system at 1000 hours?
(b) Assume in a subsequent design iteration, Component C is deleted from the system and replaced with a new component (Component F). Component F has a Weibull time-to-failure distribution ($\beta = 2.3$, $\eta = 1400$ hours, and $\gamma = 0$). Assume all the other components (A, B, D, and E) still have constant failure rates as defined above, now what is the reliability of the system at 1000 hours?

3.8 Repeat Problem 3.7 assuming that the parts are connected in parallel.

3.9 The reliability of a system is given by,

$$R(t) = \begin{cases} (1 - t/2t_0)^2, & 0 \le t \le 2t_0 \\ 0, & t > 2t_0 \end{cases}$$

where t_0 is 14 years.

(a) What is the instantaneous failure rate when $t = 8$ years?
(b) What is the instantaneous failure rate when $t = 29$ years?
(c) What is the mean time to failure ($MTTF$)?

3.10 A software program has an initial failure intensity of 20 failures/CPU hour. After 5 CPU hours of test/fix, the failure intensity dropped to 10 failures/CPU hour. How many total errors were resident in the software at time 0?

Chapter 4

Maintenance – Managing System Failure

Maintenance is the practice of keeping systems operational. Maintenance takes place before or after a failure (or both).[1] Maintenance involves functional checks, servicing, repairing or replacing of necessary devices, equipment, machinery, building infrastructure, and supporting utilities in industrial, business, governmental, and residential installations.

If a product is inexpensive enough, it is likely not maintained. For example, customers rarely take any action to maintain an inexpensive flashlight; it does not get cleaned, its battery may not be accessible, and when it fails, it is simply thrown away (even exercising its warranty, if it has one, may be more trouble than its worth). However, at some cost threshold, customers begin to take an interest in maintaining a product. A cell phone is expensive enough that its customer must obtain several years of use from it before it is replaced. In order to obtain those years of use, the customer must allow it to update its software (this is software maintenance), it has a warranty that the customer will exercise if it fails, and the customer may seek repairs for damage outside of the warranty period or terms, e.g., replacing a cracked screen.

---

[1] We will generally refer to activities discussed in this chapter as "maintenance". "Repair" represents a subset of maintenance activities that occurs after a failure. Maintenance includes repair, but also activities associated with keeping the system from failing.

Expensive complex systems are nearly always maintained because they are too expensive to discard when they fail.[2] Maintenance refers to the measures taken to keep a system in an operable condition or to repair it to an operable condition [4.1]. The amount that economies spend on maintenance is unknown, in large part because the majority of maintenance is performed in-house rather than purchased on the market. Representative numbers collected by Canada in 2016 indicated that firms spent 3.3% of Canada's GDP on repairs, which is more than double what the country spent on research and development [4.2]. Although critically important, "Maintenance lacks the glamour of innovation. It is mostly noticed in its absence" [4.2].

The term *maintainability* is used to denote the study and improvement of the ability to maintain systems, and is primarily focused on reducing the amount of time required to diagnose and repair failures.

4.1 Maintenance Levels

Maintenance is often classified based on where it is performed. Some maintenance can be performed at the system, examples include changing a flat tire on your car or an automatic software upgrade on your phone. This type of maintenance is sometimes called *organizational maintenance* (or *basic* or *O-level*) because it is performed by personnel without specialized training at the operational site by the organization that owns or operates the system. At the organizational level, maintenance is usually confined to simple activities such as inspecting, servicing, calibrating, lubricating, or adjusting equipment, as well as removal and replacement of failed components. Failed components are generally not repaired at the organizational level because repair requires equipment, facilities and expertise that are not continuously available where the system is used.

Intermediate maintenance (or *I-level*) is composed of tasks that are performed by mobile and/or fixed specialized dedicated maintenance organizations on removed parts, components, or equipment. For example, the repair of a gearbox on a wind turbine can only be

---

[2]We qualify this statement with "nearly always" because there are some expensive complex systems that are not maintainable due to the unavailability of the required resources, or the unavailability of the system, e.g., satellites. All of the discussion in this chapter assumes that the system can be repaired.

done when a crane is available at the wind farm. The crane and its supporting personnel and resources represent an intermediate maintenance organization. There are far fewer cranes than wind turbines; in fact, there are fewer cranes than wind farms, so cranes travel between wind farms and can only perform maintenance on wind turbines when they are at the farm. Wind turbines are an example of a system that is fixed in one location, in some cases, e.g., an aircraft, the aircraft can be relocated to a fixed intermediate maintenance facility where maintenance can be performed.

There are often subsystems in complex systems that cannot be repaired at either the organizational or intermediate maintenance levels. For example, the avionics electronics in an aircraft. Failed avionics can be removed and replaced at the intermediate level (maybe at the organizational level), but diagnosing and repairing a failed avionics subsystem is beyond the capabilities of either the intermediate or organizational maintenance levels. In this case the avionics subsystem will go to a *depot maintenance* (or *D-level*) facility that specializes in repairing electronic subsystems. In some cases, the depot may be operated by the original manufacturer of the subsystem (which may be located at that manufacturer's facility). In some cases, a third party that is not the user or the original manufacturer may be contracted to provide intermediate or depot level maintenance.

From a maintenance planning and modeling perspective, the primary differentiator between the maintenance levels is the cost and duration of the maintenance activity. Long downtimes waiting for maintenance to be performed lead to decreases in system availability. Table 4.1 summarizes the primary attributes of the three maintenance levels described in this section.

4.2   Maintenance Time and Maintainability

Maintainability is the probability that a failed unit will be repaired (restored to an operable state) within a given amount of time. For example, a system with a maintainability of 95% in 1 day has a 95% probability of being restored to operability within 1 day of its failure. Quantitatively assessing maintainability requires that maintenance times be understood. Maintenance time is most commonly represented by a lognormal distribution, Figure 4.1. $g(t)$ in Figure 4.1

Table 4.1. Maintenance level attributes.

Attribute	Organizational Maintenance	Intermediate Maintenance		Depot Maintenance
		Mobile	Fixed	
Location	At the system's location.	At the system's location.	At a fixed location.	At a specialized fixed location or the original equipment manufacturer's location.
Staffing	Persons without specialized maintenance training.	Persons with specialized training but limited to specific activities.	Broad set of specialized persons.	Highly specialized maintenance personnel, and engineering that designed and manufactured the original system.
Facilities/ Resources	Limited to what can be carried with the system or kept at the system's fielded location.	Limited to what can be transported to the system.	Specialized and extensive.	Equipment and facilities equivalent to the equipment used to manufacture and test the original system.

Timing	Quick turnaround.	The system needs to wait for the mobile resources to arrive on site.	The system needs to be relocated to the fixed facility.	The failed subsystem needs to be removed and sent to depot facility.
Scope	Scheduled inspections/ servicing and rapid repair. Limited to on-board diagnostics.	Significant inspections/diagnostics and component replacement, repair and overhaul.		Extensive modification or overhaul to major system components. refurbishing, reconditioning, and one-to-one replacing of defective parts, components or assemblies.

System Sustainment

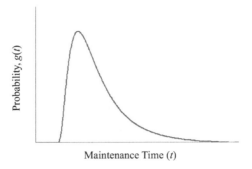

Fig. 4.1. Lognormal distribution (PDF) of maintenance times.

represents the probability of maintenance times (it is the maintenance time probability density function).[3] Figure 4.1 represents both job-to-job variability in maintenance times, maintenance personnel variability and also learning. There are also many different maintenance time measures that depend on what activities are included or excluded from the metric (see Section 4.3), but in general, any single activity or combination of activities could be represented by a distribution of maintenance times.

Mathematically, the lognormal distribution is given by

$$g(t) = \frac{1}{t\sigma\sqrt{2\pi}} e^{-\frac{1}{2}\left(\frac{\ln(t)-\mu}{\sigma}\right)^2} \tag{4.1}$$

where t is the time, μ is the mean of $\ln(t)$, the location parameter, and σ is the standard deviation of $\ln(t)$, the scale parameter.

The maintainability, $M_a(t)$, is the probability of completing maintenance in a time T, which is less than t and is given by

$$M_a(t) = \Pr(T < t) = \int_0^t g(\tau)d\tau \tag{4.2}$$

---

[3]Figure 4.1 and $g(t)$ are more commonly expressed as "repair time", but we will generalize them to "maintenance time" for our discussion in this chapter.

where $g(\tau)$ is the maintenance time probability density function. Substituting Equation (4.1) into Equation (4.2), the maintainability corresponding to lognormally distributed maintenance times becomes

$$M_a(t) = \int_0^t g(\tau)d\tau = \int_0^t \frac{1}{\tau\sigma\sqrt{2\pi}}e^{-\frac{1}{2}\left(\frac{\ln(\tau)-\mu}{\sigma}\right)^2}d\tau = \Phi\left(\frac{\ln(t)-\mu}{\sigma}\right)$$

(4.3)

where Φ is the standard normal CDF.[4] In this case the Mean Time to Repair ($MTTR$) is given by[5]

$$MTTR = \int_{-\infty}^{\infty} g(t)dt = e^{\left(\mu+\frac{\sigma^2}{2}\right)}$$

(4.4)

In general, the time-to-repair should include the time to diagnose, disassemble, and transport the failed unit to a place it can be repaired; obtain replacement parts and other necessary materials; make the repair; perform necessary functional testing; reassemble the unit; and verify and test the unit in the field. The $MTTR$ of a system of N components can be calculated using [4.3]

$$MTTR = \frac{\sum_{i=1}^{N} \lambda_i MTTR_i}{\sum_{i=1}^{N} \lambda_i}$$

(4.5)

where λ_i is the failure rate of the ith component of the system.

The repair rate (repairs per unit time) is given by

$$\mu_{rr}(t) = \frac{g(t)}{1 - M_a(t)}$$

(4.6)

Serviceability is used to articulate the degree of difficulty of repair. Maintainability (or repairability) infers the probability of maintaining in a specified time, while serviceability is a system design characteristic that may correlate to maintenance time.[6]

---

[4]The standard normal CDF is given by $\Phi(x) = \frac{1}{\sqrt{2\pi}}\int_{-\infty}^{x} e^{-t^2/2}dt = \frac{1}{2}\left[1 + \text{erf}\left(\frac{x}{\sqrt{2}}\right)\right]$.

[5]Note, the units on $MTTR$ will be the same as the units on t since μ is the mean of $\ln(t)$.

[6]There are several alternative definitions of serviceability in various domains. For example, serviceability for structures refers to the conditions under which a building is still considered useful. For computers, it is a synonym for supportability.

4.3 Common Maintenance Measures

There are many time measures associated with maintenance that can be computed and tracked by organizations. Some of the most relevant time measures are summarized in Table 4.2. For context, many of the time measures in Table 4.2 are used to calculate various availability metrics for the system (see Chapter 5).

Table 4.2. Selected maintenance time measures.

Symbol	Name	Content
$MTBF$	Mean time between failures	Mean time between corrective maintenance activities.
$MTTR$ ($\overline{M}_{ct}$)	Mean time to repair (Mean corrective maintenance time)	Corrective maintenance (as a result of failure): failure detection, diagnosis (fault isolation), disassembly, repair, reassembly, verification, etc.
$MTBM$	Mean time between maintenance	Mean time between all (corrective and preventative) maintenance activities.
$MTPM$	Mean time to perform preventative maintenance	
$\overline{M}$	Mean active maintenance time	Corrective and preventative maintenance (weighted sum of $\overline{M}_{ct}$ and $\overline{M}_{pt}$).
MDT	Mean maintenance downtime	$\overline{M}$ with LDT and ADT included
$\overline{M}_{pt}$	Mean preventative maintenance time	Preventative maintenance: scheduled maintenance, periodic inspection, servicing, calibration, overhaul, etc. Can overlap with $\overline{M}_{ct}$ and operational time.
LDT	Logistics delay time	Time spent waiting for spares, test equipment, and/or facilities; transportation time.
ADT	Administrative delay time	Time spent waiting for personnel assignments, prioritization, organizational delays, etc.
MSD	Mean supply delay	$LDT + ADT$

As an example, consider a system with the following characteristics ("op hours" = operational hours):

- Operational cycle = 2000 op hours/year
- Support life = 5 years
- Failures that require corrective maintenance = 2/year
- Repair time per failure = 40 op hours ($MTTR$)
- Preventative maintenance activities = 1/year
- Preventative maintenance time per preventative maintenance action = 8 op hours ($MTPM$)
- Average wait time for repair materials for corrective maintenance = 10 op hours (LDT)

From the given information the following quantities can be calculated:

$$\text{Total number of maintenance actions} = (2)(5) + (1)(5) = 15 \quad (4.7a)$$

$$\overline{M} = \frac{(40)(2)(5) + (8)(1)(5)}{15} = 29.333 \, \text{op hours} \quad (4.7b)$$

$$MDT = \frac{(40 + 10)(2)(5) + (8)(1)(5)}{15} = 36 \, \text{op hours} \quad (4.7c)$$

$$MTBF = \frac{(5)(2000)}{(2)(5)} = 1000 \, \text{op hours} \quad (4.7d)$$

$$\text{Total operational cycle} = (5)(2000) = 10,000 \, \text{op hours} \quad (4.7e)$$

$$\text{Total downtime} = (15)(36) = 540 \, \text{op hours} \quad (4.7f)$$

$$\text{Total uptime} = 10,000 - 540 = 9460 \, \text{op hours} \quad (4.7g)$$

$$MTBM = \frac{9460}{15} = 630.667 \, \text{op hours} \quad (4.7h)$$

The simple maintenance metrics articulated in this section are far from exhaustive. Availability metrics will be addressed in Chapter 5. Additional metrics that measure planned and preventative maintenance versus corrective maintenance are also relevant, but will be addressed in the sections that follow. There are also inventory metrics that will be discussed in Chapter 6.

4.4   Corrective Maintenance

"Corrective maintenance is a maintenance task performed to identify, isolate, and rectify a fault so that the failed equipment, machine, or system can be restored to an operational condition within the tolerances or limits established for in-service operations" [4.4]. Corrective maintenance (also known as "break-fix", "run-to-failure", or "breakdown" maintenance) is initiated after a system fails.[7] Corrective maintenance can result in a repair of the system, a restart of the system, or a replacement of the system. Corrective maintenance has two forms: "immediate" (resolution begins immediately upon failure) and "deferred" (resolution is delayed for some period of time based on maintenance policies).

The cost of corrective maintenance is the product of the number of system failures that have to be resolved and the cost of resolving them. If the system has a constant failure rate, the reliability of the system was derived previously in Equation (3.17) as,

$$R(t) = e^{-\lambda t}$$

where t is time and λ is the failure rate. The mean time between failure ($MTBF$) for this system is $1/\lambda$. Suppose, for simplicity, the failures of this system are resolved instantaneous at a maintenance cost of \$1000/failure. If we wish to support the system for 20 years and the units of λ are failures/year, how much will it cost? Assuming that the discount rate on money is zero, this is a trivial calculation:

$$Total\ Cost = 1000(20\lambda) \tag{4.8}$$

The term in parentheses in Equation (4.8) is the total number of failures in 20 years. If $\lambda = 2$ failures per year, the *Total Cost* is \$40,000. If we include a cost of money (see Appendix A), using a discrete discount rate (r), the solution becomes a sum, because each

---

[7]Note, in some domains, corrective maintenance refers to maintenance that occurs prior to failure as the result of an inspection activity, e.g., this is the case in some equipment/machinery focused maintenance domains. In these domains, what we are calling corrective maintenance may be referred to as "emergency maintenance".

maintenance event has a different cost in year 0 dollars,

$$Total\ Cost = \sum_{i=1}^{20\lambda} \frac{1000}{(1+r)^{\frac{i}{2}}} \tag{4.9}$$

where $i/2$ is the event date in years.[8] If we assume $r = 8\%$/year, the *Total Cost* will be \$20,021.47 in year 0 dollars.

In reality, the actual event dates in the example presented above are not known (they do not happen at exactly *MTBF* intervals), rather the time-to-failures are represented by a failure distribution. The failure distribution can be sampled to capture a sequence of failure events whose costs can be summed (see Section C.2.2).

4.4.1   *Spare parts*

Spare parts are parts that are identical to original parts that are available to be used as replacements for the original parts if the original parts fail. Spare parts exist because the availability of a system is important to its owner or users. *Availability* (see Chapter 5) is the ability of a service or a system to be functional when it is requested for use or operation. Availability is a function of an item's reliability (how often it fails) and maintainability (how efficiently it can be restored when it does fail). Having your car unavailable to you because no spare tire exists is a problem. If you run an airline, having an airplane unavailable to carry passengers because a spare part does not exist or is in the wrong location is a problem that results in a loss of revenue.

There are numerous challenging issues that arise with spare parts. The most obvious question is, how many spares do you need to have? There is no need to purchase or manufacture 1000 spares if you will only need 200 to keep the system operational (available) at the required rate for the required time period. The calculation of the quantity of spares is addressed in the remainder of this section. The second problem is, when are you going to need the spares? The number of spares I need is a function of time (or miles, or other

---

[8]The $i/2$ assumes that $\lambda = 2$ and the failures are uniformly distributed throughout the year.

accumulated environmental stresses); as systems age, the number of spares they need as a fraction of time may increase. If possible, spares should be purchased over time rather than all at once at the beginning of the life cycle of the system. The disadvantages of purchasing all the spares up front are the cost of money, holding costs and finite shelf life. However, in some cases the procurement life of the spares (see Chapter 7) may preclude the purchase of spares over time.

The issues with spares extend beyond quantity and time. Spares also have to be stored somewhere. They should be distributed to the places where the systems will be when they fail or, more specifically, where the failed system can be repaired. Is a spare tire more useful in your garage or in the trunk of your car? On the other hand, does it make sense to carry a spare transmission in the trunk of your car? Probably not – transmissions rarely fail and a transmission cannot be installed into the car on the side of the road. The process of acquiring and distributing spares is the realm of inventory modeling, which will be discussed in Chapter 6.

4.4.1.1 *Calculating the number of spares needed*

The most important question associated with spares is usually: "how many will I need?" Knowing the quantity of spares is central to planning, i.e., budgeting future money and potentially appropriate storage facilities to accommodate the spares that are necessary to keep a system operational. If there was no uncertainty in the failure rate associated with a system, this would be a trivial calculation. In the simple example described in the introduction to this section, 20λ in Equation (4.8) is the number of the number of "spares" needed to support the system for 20 years (if $\lambda = 2$ failures/year then 40 spares are necessary). However, there are numerous implicit assumptions in this simple calculation:

- The spares used are exactly as-good-as-new. In other words, the spare parts have failure rates that are identical to the original parts (not better, not worse).[9]

---

[9]In reality, the reliability of a spare could be better than the original (e.g., it is possible that the technology or materials have improved since the original was fielded), or worse (e.g., if the spare degrades in storage, or design/technology

- The parts are non-repairable, i.e., the spares permanently replace the failed part.[10]
- There are no uncertainties, i.e., the failure rate is exactly λ every year.

Let's consider the third assumption in detail. Unfortunately, everything is uncertain, which means that just asking "how many spares will be needed" isn't concise enough. We also must specify the desired confidence that we will have enough spares. In addition, we must specify more carefully how we plan to use the spares. Generally, spares can be used in one of two ways (Figure 4.2):

- *Permanent*: the spare permanently takes the place of the original part, i.e., the original failed part is disposed of and the spare takes its place until the end of support of the system or the spare's failure, whichever comes first.

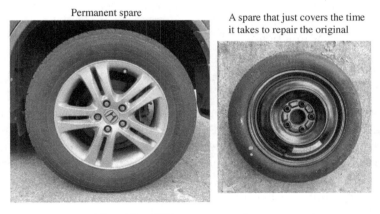

Fig. 4.2.   Different types of spares.

---

changes could result in parts that are, for example, less expensive, but not as reliable).

[10]Items for which spares exist are generally classified into non-repairable and repairable. A repairable item is one that, upon removal from operation due to a preventative replacement or failure, is sent to a repair or reconditioning facility, where it is restored to an operational state (nominally, good-as-new). Non-repairable items are discarded once they have been removed from operation (or possibly recycled to recover materials).

- *Compensate for the failed item*: the spare only replaces the failed item for the period of time during which the original item is repaired, or possible while a more appropriate replacement part is acquired.

The following calculation is general, but the interpretation of time in the calculation depends on the way the spare is used.

If we assume a constant failure rate λ, the expected number of failures (expected number of spares needed) in time t is λt (this result will be rigorously determined in Section 4.4.1.2), but if I buy λt spares, what is my confidence that I have bought enough spares to last until t? Equation (3.17) gives the reliability for a constant failure rate. This is the probability of exactly 0 failures in time t, i.e., $\Pr(0) = R(t) = e^{-\lambda t}$. If I knew the probability of k or fewer failures in time t, then k would be the number of spares that corresponds to that probability.

To develop a relation that relates the number of spares (k) to the probability that those spares will be sufficient, we start with the binomial distribution, which is the probability of getting exactly i successes in N independent Bernoulli trials,

$$\Pr(i; N, p) = \frac{N!}{i!(N-i)!} p^i (1-p)^{N-i} \qquad (4.10)$$

where p is the probability of a successful outcome for a single trial. Equation (4.10) is the probability mass function of the binomial distribution. A common example is flipping coins where p is the probability of getting a head on one flip, and there are only two possible outcomes (heads or tails), so the system is binomial. If the coin is flipped N times, Equation (4.10) gives the probability of getting exactly i heads.

The cumulative distribution function is the sum of the individual probabilities predicted by Equation (4.10),

$$\Pr(X \leq k; N, p) = \sum_{i=0}^{k} \frac{N!}{i!(N-i)!} p^i (1-p)^{N-i} \qquad (4.11)$$

The mean of the binomial distribution is Np, i.e., if a coin is flipped $N = 100$ times and the probability of getting a head on a single toss is $p = 0.5$, then the expected number of heads is $(100)(0.5) = 50$. If the failure rate (mean rate of occurrence of a failure) is λ, then λt is the expected number of failures in t, so $Np = \lambda t$ and when N is large, we get[11]

$$\Pr(X \leq k) \cong \sum_{i=0}^{k} \frac{(\lambda t)^i}{i!} e^{-\lambda t} \qquad (4.12)$$

Note, when $k = 0$, Equation (4.12) gives $e^{-\lambda t}$, which is the reliability given by Equation (3.17). For n systems that have to be supported (not redundant systems), we can replace λt with $n\lambda t$,

$$\Pr(X \leq k) \cong \sum_{i=0}^{k} \frac{(n\lambda t)^i}{i!} e^{-n\lambda t} \qquad (4.13)$$

where

k is the number of spares,

n is the number of unduplicated (in series, not redundant) units in service,

λ is the mean failure rate of the unit or the average number of maintenance events expected to occur in time t,

t is the time interval, and

$\Pr(X \leq k)$ is the probability that k is enough spares or the probability that a spare will be available when needed (this is known as the "protection level", "probability of sufficiency", or "safety factor").

Solving Equation (4.13) for k gives the number of spares needed. However, the time interval (t) in Equation (4.13) can be interpreted several ways. If the spares are permanent than t is the total time that the system needs to be supported. Conversely, if the spares are only required to support the system while the original failed item is being repaired, then t is the time-to-repair the original item.

If λ represents a constant failure rate Equation (4.13) is a good approximation. What if the failure rate is not constant? How can

---

[11]To get from Equation (4.11) to (4.12): (1) replace p in Equation (4.11) with $\lambda t/N$; (2) take the natural log of both sides of the resulting equation; and (3) expand with a Taylor series expansion and assume N is large.

we use Equation (4.13)? For help we go to Palm's Theorem [4.5]. Palm's theorem essentially states that if the demand for an item is given by a Poisson process with a mean of λ per unit time, and if the repair time for each failed unit is independent and identically distributed according to any distribution with a repair time t, then the steady-state probability distribution for the number of units in repair has a Poisson distribution with a mean of λt. The proof of this is beyond the scope of this text, but the implication to us is that Equation (4.13) will work for any failure distribution as long as we can determine the mean failure rate (λ) for the distribution.

As an example, consider a bus that is intended to operate for 200,000 miles per year. Assume that the failure of the component of interest follows a Weibull distribution with $\beta = 2$, $\eta = 74,000$ miles and $\gamma = 0$. How many spares are needed to achieve a 90% protection level? To solve this, we use Equation (4.13) with $n = 1$, $t = 200,000$, and the mean of the Weibull distribution, which is $\eta\Gamma(1 + 1/\beta) = 65,581$ miles, so $\lambda = 1/\text{mean} = 1.5248 \times 10^{-5}$ failures/mile. Equation (4.13) gives 0.911 when $k = 5$, so the bus requires 5 or more spares in order to have a 90% confidence that it has enough spares. What about 5 buses? Now set $n = 5$ and solve, $k = 20$ gives a protection level of 0.906 – note that k is not $(5)(5) = 25$, but less. This is because all 5 buses can all draw spares from the same pool; if each bus had a segregated pool of spares that only it could draw from, then the answer would have been 25.

For a sufficiently large value of $n\lambda t$ (i.e., > 10), a normal distribution with a mean of $n\lambda t$ and a variance of $n\lambda t$, is a good approximation to the Poisson distribution. In this case the number of spares needed can be calculated from [4.6],

$$k \cong \lceil n\lambda t + z\sqrt{n\lambda t}\rceil \qquad (4.14)$$

where z is the number of standard deviations from the mean of a standard normal distribution.[12] This approximation is independent of the underlying failure distribution. For the example given above, with $n = 1$, k from Equation (4.14) is 6, and with $n = 5$ it is 21.

---

[12]Single-sided z-score. $z = \text{NORMINV}(PL,0,1)$ in Excel, where PL is the desired protection level.

4.4.1.2   *Warranty, renewal functions and the number of spares needed*

When a warranty is defined for a product, it is important to determine how much it will cost to fund the warranty so that the appropriate amount of money can be collected from the customers as part of the price of the product. In order to determine the cost of the warranty we must predict the number of failures the product will have during the warranty period.

Renewal functions are another way of estimating the quantity of spares. A renewal function gives the expected number of failures of a system in an interval. The number of renewals and k (the number of spares) are not the same thing. k is the number of spares necessary to satisfy a specified confidence that you have enough spares to last t (i.e., $\Pr(X \leq k)$ in Equation (4.13)). The number of renewals is the expected number of spares needed to last for t. Renewal functions are commonly used to estimate warranty reserve funds for a warranty period of t and to estimate maintenance budgets, but if one wants to know how many spares are necessary to satisfy a particular confidence level then a treatment like that in Equation (4.13) or (4.14) is necessary.

Consider a product that is placed in operation at time 0. When the product fails at some later time it is immediately replaced with a new version of the product (a spare) that has a reliability identical to the original unit at time 0. When the spare product fails after a time it is similarly replaced by a good-as-new version of the product. The expected number of failures and associated renewals per product instance within a population of the product in the interval $(0,t]$ is denoted by a renewal function, $M(t)$. If we account for only the first failure, $M(t) = F(t) = 1 - R(t)$, where $F(t)$ is the unreliability and $R(t)$ is the reliability. This estimation of $M(t)$ assumes that repaired or replaced products never fail. The difference between $M(t)$ and $F(t)$ is that $M(t)$ accounts for more than the first failure, including the possibility that the repaired or replaced product may fail again during the warranty period.[13]

---

[13]The renewal function $M(t)$ is sometimes referred to as the mean cumulative function (MCF).

The derivation of the renewal function is beyond the scope of this text, however, it can be shown that the Laplace transform of the renewal function is given by

$$\hat{M}(s) = \frac{1}{s}\left[\frac{\hat{f}(s)}{1 - \hat{f}(s)}\right] \tag{4.15}$$

where $\hat{f}(s)$ is the Laplace transform of the time-to-failure PDF.

For a constant failure rate of λ, the $f(t)$ is given by Equation (3.15). The Laplace transform of $f(t)$ is

$$\hat{f}(s) = \frac{\lambda}{s + \lambda} \tag{4.16}$$

Substituting Equation (4.16) into Equation (4.15) gives

$$\hat{M}(s) = \frac{\lambda}{(s + \lambda)s\left(1 - \frac{\lambda}{s+\lambda}\right)} = \frac{\lambda}{s^2} \tag{4.17}$$

and taking the inverse Laplace transform,

$$M(t) = \lambda t \tag{4.18}$$

If, for example, a system with a constant failure rate of 1×10^{-5} failures per hour of continuous operation has a 1-year warranty, and if 10,000 of these systems are fielded, what is the expected number of legitimate warranty claims during the warranty period? From Equation (4.18), $M(t) = (1 \times 10^{-5})(24)(365) = 0.0876$ expected failures per unit. So the expected number of claims is $(0.0876)(10,000) = 876$ claims. Note, $F(t) = 1 - e^{-\lambda t} = 0.0839$, which is smaller than 0.0876.

Equation (4.15) can be used to determine the renewal function for other failure distributions, however, unlike the exponential distribution, for most other distributions $M(t)$ does not have a simple closed form. There are several approximations for renewal functions. The following non-parametric renewal function estimation for large t ($t >> \mu$) is commonly used [4.7]:

$$M(t) \cong \frac{t}{\mu} + \frac{\sigma^2}{2\mu^2} - \frac{1}{2} \tag{4.19}$$

where μ and σ^2 are the mean and variance of the failure distribution given by

$$\mu = -\frac{d\hat{f}(s)}{ds} \text{ and } \sigma^2 = \frac{d^2\hat{f}(s)}{ds^2} - \mu^2 \qquad (4.20)$$

both evaluated at $s = 0$.

Equations (4.19) and (4.20) are valid for any distribution. For example, for exponentially distributed failures, $\mu = 1/\lambda$ (the *MTBF*) and $\sigma^2 = 1/\lambda^2$, which from Equation (4.19) gives, $M(t) = \lambda t$, which is the same result derived from Equation (4.18). For a 2-parameter Weibull distribution ($\gamma = 0$), the mean and variance are given by

$$\mu = \eta\Gamma\left(1 + \frac{1}{\beta}\right) \text{ and } \sigma^2 = \eta^2\left[\Gamma\left(1 + \frac{2}{\beta}\right) - \Gamma^2\left(1 + \frac{1}{\beta}\right)\right] \qquad (4.21)$$

where $\Gamma(\ )$ denotes a gamma function. Using Equations (4.21) and (4.19), an approximation to the renewal function for a Weibull distribution can be found.

To illustrate the analysis of maintenance costs, consider a bus that is intended to operate for 200,000 miles per year. Reliability analysis indicates that the failure of a critical component follows an exponential distribution with a failure rate of $\lambda = 1.4 \times 10^{-5}$ failures/mile.[14] Assume that the replacement component is "as-good-as-new" and that the failure mechanism only accumulates damage while the bus is operating (not while it is being repaired). What is the expected maintenance cost for one bus, for 1 year?

The component failures follow an exponential distribution, so we can use Equation (4.18) to estimate the number of renewals in a 1 year (200,000 mile) period. Using Equation (4.18) we get $M(t = 200{,}000) = 2.8$ renewals/year (repairs in this case).

How many spares do we need to have a 90% confidence that we have enough spares for one bus for 1 year? 2.8 is the expected number of spares (per bus per year). To solve this problem, we need to use Equation (4.13) with $n = 1$ (one bus). When $k = 3$ the confidence level is $\Pr(x \leq k) = 0.69$; to obtain a confidence level greater than

---

[14]Note, everything in this illustration is in miles rather than time. Mileage can be converted to time if desired, but it is not necessary to do so. We are also assuming that all maintenance is via component replacement, i.e., there is no component repair.

0.9, $k = 5$ spares have to be used, $\Pr(x \leq k) = 0.93$ in this case. We could also obtain this result using discrete-event simulation, see Section C.2.3.

The field of warranty cost analysis (e.g., see [4.8]) focuses on the determination of $M(t)$ for complex warranty periods, e.g., 2-dimensional warranties such as 3 years or 36,000 miles, whichever comes first.

4.4.1.3 *Alternating renewal processes*

In the bus example in the last section, what if we assume that it takes 5 days (2740 miles of lost bus usage) each time the component must be replaced when it fails and that the component of interest only accumulates damage when the bus is being used (not while it is being repaired)? The 2.8 renewals/year we computed would be the correct number of repairs if the relevant failure mechanism accumulated damage continuously over calendar time, but because it only accumulates damage when it is operating, 2.8 is too large. The time (miles) to perform the corrective maintenance is not zero (the calculation in the previous section implicitly assumes it is zero, i.e., it assumes the bus is fixed instantaneously on failure, which it is not).

The renewal process that was used to generate Equation (4.15) assumes that the component of interest is non-repairable (it simply has to be replaced), and that the process of replacing it takes a very short period of time (much shorter than the time between replacements). An alternating renewal process is composed of two processes that do not occur simultaneously, but rather, alternate in time. For the bus example with a non-zero repair time, the two processes are a time-to-failure of the system and a time-to-repair (maintenance recovery) of the system, each of which is governed by its own probability distribution.

For an alternating renewal process, Equation (4.15) becomes

$$\hat{M}(s) = \frac{1}{s} \left[\frac{\hat{f}(s)\hat{g}(s)}{1 - \hat{f}(s)\hat{g}(s)} \right] \tag{4.22}$$

where $\hat{f}(s)$ is the Laplace transform of the time-to-failure PDF, and, for the bus example, $\hat{g}(s)$ is the Laplace transform of the time-to-repair PDF.

As an example of an alternating renewal problem, consider again the bus example from the last section with $\lambda = 1.4 \times 10^{-5}$ failures/mile. Assume that it takes 5 days (2740 miles of lost bus usage) and costs \$5,000 each time the component must be replaced when it fails. In this case, how many renewals are there per year? To solve this using Equation (4.22) we would substitute Equation (4.16) for $\hat{f}(s)$ and,

$$g(t) = \delta(t - 2740) \tag{4.23a}$$

$$\hat{g}(s) = e^{-2740s} \tag{4.23b}$$

We aren't equipped to solve this problem in this text (see [4.9] for a solution), but we can approximate the solution by adjusting the failure rate in the following way,

$$\lambda_{\text{modified}} = \frac{1}{1/\lambda_{\text{original}} + 2740} = 1.348 \times 10^{-5} \frac{\text{failures}}{\text{mile}} \tag{4.24}$$

Equation (4.24) effectively extends the $MTBF$ ($1/\lambda_{\text{original}}$), by the maintenance duration. Using the new value of λ, $M(t = 200{,}000) = 2.697$ renewals. Note, this problem can also be solved using discrete-event simulation, see Section C.2.3.

Based on only corrective maintenance, the annual maintenance cost for a bus is given by

$$Cost_{\text{annual}} = c_f M(t) \tag{4.25}$$

where c_f is the cost per maintenance event. For the bus problem, from Equation (4.25) with $c_f = \$5000$, the annual maintenance cost per bus is \$13,483.

4.5 Preventative Maintenance

Preventative maintenance potentially avoids more expensive corrective maintenance. Preventative maintenance (also called "scheduled" or "fixed-interval" maintenance) is performed at some predetermine interval, which can be measured in time or other quantities that correlate to damage accumulation (miles, takeoffs/landings, on/off cycles, etc.). Generally preventative maintenance intervals are conservative, i.e., they are selected so that failure is avoided. For example, if you change the oil in your car every 3000 miles, you are performing preventative maintenance. Corrective maintenance is generally more costly because it occurs at unplanned times potentially making the logistics of repair more difficult (i.e., at a place or time where the resources to perform the repair may not be available), and it may cause collateral damage to other system components.[15]

Like corrective maintenance, preventative maintenance can be not-as-good-as-new, good-as-new, or better-than-new. Preventative replacement is a form of preventative maintenance that replaces a component with a new component that is good-as-new. The objective of preventative maintenance analysis is to determine the optimum interval at which to perform the maintenance. If preventative maintenance is performed too often, then a great deal of money is spent performing the preventative maintenance, alternatively, if the interval between preventative maintenance actions is too long, then the system might experience a large number of corrective maintenance events. Setting aside safety arguments, there is potentially an optimum preventative maintenance interval that is not so conservative that it does not allow for some corrective maintenance.

Let's consider a couple of trivial cases. Suppose I have a system that costs $c_f = \$1000$ to repair on failure (corrective maintenance cost), and $c_p = \$100$ to perform preventative maintenance (assume that preventative maintenance restores the system to good-as-new). For our first exercise, assume that the reliability is of the system is exactly 450 hours. When should preventative maintenance be done? The answer is of course trivial, it should be done an instant before 450 hours; this avoids all corrective maintenance and minimizes the number of preventative maintenance actions. Usually we don't know when the actual failure will be, and we need to find a preventative maintenance interval that provides the optimum balance between being too conservative and performing too much corrective maintenance.

---

[15]Collateral damage is damage to other components or subsystems that is the direct result of the failure of a different portion of the system. For example, the failure of a tire when a car is traveling at 60 miles per hour could result in collateral damage to brakes, axels and other car subsystems.

There are several approaches to solving this problem depending on what preventative maintenance policy we adopt.

4.5.1 *Constant-interval replacement*

If I assume constant-interval replacement of a component independent of whether the component failed and was replaced within the interval, then

$$Cost(t_p) = \frac{\text{Total expected replacement cost in the interval}}{\text{Length of the interval}}$$

$$= \frac{1c_p + M(t_p)c_f}{t_p} \tag{4.26}$$

where we have assumed that the interval length is t_p and $M(t_p)$ is the number of renewals in the interval $(0, t_p]$.

Constant-interval replacement is not very realistic. If a component failed within the interval and was replaced in the interval, then it is likely that that component will be exempted from maintenance until t_p past the point when corrective maintenance was done.

4.5.2 *Replacement at a predetermined age*

A more realistic preventative maintenance scheme is the replacement of a component upon failure or when it reaches an age of t_p whichever comes first. To assess the cost of a system with this combination of corrective and preventative maintenance, we define a maintenance cycle length, which is the length of time between maintenance events (corrective or preventative). In terms of this maintenance cycle length, the total maintenance cost per unit time is given by [4.10]

$$Cost(t_p) = \frac{\text{Total expected replacement cost}}{\text{Expected maintenance cycle length}}$$

$$= \frac{R(t_p)c_p + [1 - R(t_p)]c_f}{R(t_p)t_p + \int_0^{t_p} tf(t)dt} = \frac{R(t_p)c_p + [1 - R(t_p)]c_f}{\int_0^{t_p} R(t)dt} \tag{4.27}$$

where

t_p is the preventative maintenance time,

c_p is the preventative maintenance cost,

c_f is the corrective (on failure) maintenance cost,

$R(t)$ is the reliability at time t,

$1\text{-}R(t)$ is the unreliability at time t, and

$f(t)$ is the PDF of the failure distribution.

The optimum maintenance interval (t_p), is determined by minimizing value of $Cost(t_p)$, i.e., determining the value of t_p that satisfies $dCost(t_p)/dt_p = 0$. For the bus problem with a constant failure rate described in Section 4.4.1.2, $Cost\ (t_p)$ is minimized when $t_p = \infty$, why? An exponential distribution is memoryless, i.e., the failure rate is constant and independent of the age of the system or whether preventative maintenance has been done. In order for preventative maintenance to make sense there must be an increasing failure rate over time, i.e., the system has to age.

To demonstrate preventative maintenance, let's change the example from Section 4.4.1.2. Assume that the failure of the component of interest follows a Weibull distribution with $\beta = 2$ (an increasing failure rate), $\eta = 74{,}000$ miles and $\gamma = 0$. Assuming just corrective maintenance, and using Equations (4.19)–(4.21) with the addition of 2740 miles to μ, the $M(t = 200{,}000) = 2.553$.[16] Let's assume that a scheduled preventative replacement task that takes 1 day (550 miles of lost usage) and costs \$2050. In this case $dCost(t_p)/dt_p = 0$ when $t_p = 65{,}500$ miles (solved numerically ignoring the time to perform maintenance). At $t_p = 65{,}500$ miles, Equation (4.27) gives $Cost(t_p) = \$0.07056/\text{mile}$. Using discrete-event simulation (see Appendix C), the average number of corrective maintenance events per year per bus is 1.976/year and the average number of preventative maintenance events per year per bus is 1.498/year. The annual cost per bus is given by ($c_f = \$5000$, $c_p = \$2050$),

$$Cost_{\text{annual}} = c_f(1.976) + c_p(1.498) = \$12{,}951 \qquad (4.28a)$$

$$Cost_{\text{annual}} = Cost(t_p)(200{,}000) = \$14{,}111 \qquad (4.28b)$$

Equations (4.28a) and (4.28b) don't result in the same cost. They don't match because the simulation, Equation (4.28a), which is more

---

[16] To get this, calculate $M(t)$ using Equation (4.19), but add 2740 miles to μ.

accurate, accommodates incomplete maintenance cycles (for which the incomplete portion is free).[17]

4.6 Predictive Maintenance

Preventative maintenance occurs on some pre-determined schedule, e.g., every 65,500 miles in the example in Section 4.5.2. Predictive maintenance occurs when the system needs maintenance based on some type of prediction. The source of the prediction differentiates the various approaches to predictive maintenance.

A common form of predictive maintenance is called condition-based maintenance (CBM). CBM uses real-time data from the system to observe the system's state (condition monitoring) and thus determine its health.[18] CBM then allows action to be taken only when maintenance is necessary [4.11]. CBM allows minimization of the remaining useful life (RUL) of the system component that would be thrown away by implementing fixed-interval maintenance (preventative) policies and avoidance of failures that accompany purely corrective maintenance policies. When the condition of the system (CBM) is coupled with the expected future environmental stress conditions the approach is referred to as prognostics and health management (PHM). In the case of PHM, predictive maintenance is performed based on the prediction of a RUL. The RUL provides a time period prior to failure in which maintenance can be scheduled to minimize the interruption to system operation.[19]

---

[17]If the length (in miles) of the problem is increased, the two models will converge to the same cost.

[18]CBM is performed as a result of some combination of sensor inputs from which one can infer the system's condition and decide that maintenance is or is not needed. "Inspection" is often performed at a frequency that is independent of the condition of the system. An inspection may produce information that helps to determine the condition of the system, i.e., it represents a "sensor" used to support CBM.

[19]For example, if an airline had a 24-hour RUL prediction (assume there is no uncertainty in this prediction), the aircraft could be rerouted to ensure that it was at an airport that has the appropriate maintenance resources between midnight and 6 a.m. tomorrow morning to obtain the required maintenance without interrupting any flight schedules.

Both CBM and PHM, however, are costly to implement and maintain, are they worth it? The economics of predictive maintenance includes predicting the return-on-investment (ROI) associated with investing in predictive maintenance; and optimizing when to act (and what action to take) when a predicted RUL (including its associated uncertainties) is obtained.

A cost avoidance (see Section 8.5.2) ROI for PHM can be calculated using [4.12]

$$ROI = \frac{\text{Cost Avoided-Investment}}{\text{Investment}} = \frac{C_u - C_{PHM}}{I_{PHM}} \qquad (4.29)$$

where

C_u is the life-cycle cost of the system managed using unscheduled (corrective) maintenance,

C_{PHM} is the life-cycle cost of the system when managed using a PHM (predictive) maintenance approach, and

I_{PHM} is the investment in PHM when the system is managed using a PHM (predictive) maintenance approach.

To illustrate an ROI analysis for PHM, consider the bus example from the previous two sections. As part of the business case for the inclusion of PHM into a particular subsystem in the bus, its ROI has to be assessed. Assume the following:

- The system will fail 3 times per year
- Without PHM, all 3 failures will result in unscheduled maintenance actions
- With PHM, 2 out of the 3 failures per year can be converted from unscheduled corrective to scheduled maintenance actions (the third will still result in an unscheduled maintenance action)
- The cost of an unscheduled maintenance action is $5000 and takes 5 days of downtime
- The cost of a preventative maintenance action is $1000 (all repairs, no spares) and takes half a day of downtime
- The recurring cost (per system instance) of putting PHM into the system is $20,000
- In addition, you have to pay $2000 per year (per system instance) to maintain the infrastructure necessary to support the PHM in the system
- The bus has to be supported for 25 years

We wish to calculate the ROI of the investment in PHM relative to performing all unscheduled (corrective) maintenance. First, consider a case where the discount rate is 0. The analysis is simple in this case,

$$C_u = (25)(3)(\$5000) = \$375,000$$

$$C_{PHM} = (25)[(1)(\$5000) + (2)(\$1000)] = \$175,000$$

$$I_{PHM} = \$20,000 + (25)(\$2000) = \$70,000$$

$$ROI = \frac{375,000 - 175,000}{70,000} = 2.86$$

If the discount rate is non-zero, the calculation becomes more involved; for a 5%/year discount rate (see Appendix A for background on discounting cash flows) the solution becomes,[20]

$$C_u = \sum_{i=1}^{25} \frac{(3)(\$5000)}{(1+0.05)^i} = (3)(\$5000)\frac{(1+0.05)^{25} - 1}{(0.05)(1+0.05)^{25}} = \$211,409$$

$$C_{PHM} = \sum_{i=1}^{25} \frac{(1)(\$5000) + (2)(\$1000)}{(1+0.05)^i} = \$98,658$$

$$I_{PHM} = \$20,000 + \sum_{i=1}^{25} \frac{\$2000}{(1+0.05)^i} = \$48,188$$

$$ROI = \frac{211,409 - 98,658}{48,188} = 2.34$$

In reality, the ROI calculation associated with adding health management to a system is more complex than the simple analysis provided above. For example, predictive maintenance (e.g., PHM), will result in a combination of repairs and replacements with spares. Since the health management system will tell the maintainer to take action prior to the actual failure, some remaining life in the original component will be disposed of, which could eventually translate into the need for a greater number of spares.

---

[20] There are several implicit assumptions in this analysis including that all charges for maintenance occur at the end of the year (end-of-year convention), that the $20,000 investment in PHM occurs at the beginning of year 1, and discrete annual discounting. In this case the values of C_u and C_{PHM} are both year 0 present values.

A positive or negative ROI does not make or break a business case, but, being able to assess an ROI is part of making a business case to management or to a customer.

Predictive maintenance options that optimize when to take action after an RUL is predicted are discussed in Section 10.4.3.

4.7   Reliability-Centered Maintenance (RCM)

RCM is a process that evaluates the system in terms of its possible futures, the consequences of failures, and the maintenance procedures that should be performed. RCM ensures maintenance tasks are performed in an efficient, cost-effective, reliable, and safe manner [4.13]. Maintenance tasks may be preventive, predictive, and/or involve nondestructive inspections. The purpose of RCM is to ensure maintenance and inspection tasks are centered around improving the reliability and safety of equipment.

Reliability-centered maintenance is an engineering framework that enables the definition of a complete maintenance regimen. SAE JA1011 sets the seven questions below, worked through in the order that they are listed [4.14]:

1. What is the system supposed to do and what are its associated performance standards?
2. In what ways can the system fail to provide its required functions?
3. What are the events that cause each failure?
4. What happens when each failure occurs?
5. In what way does each failure matter?
6. What tasks or actions can be performed proactively to prevent, or to diminish to a satisfactory degree, the consequences of the failure?
7. What must be done if a suitable preventive task cannot be found?

Essentially, RCM is the process you use to determine if and when you should apply predictive maintenance.

4.8   Contract Maintenance

Under a maintenance contract, the owner of an asset pays a third party to perform contractually-specified maintenance on the asset rather than performing it themselves. On the surface this may seem simple, but for critical systems maintenance contracts can be complex. Contracts must specify what maintenance the third party is responsible for (and possibly more importantly, what they are not responsible for), timing (what timeframe the third party has to perform the maintenance in), duration of the contract, and who pays for and holds spare parts.

There are several concepts that are relevant to this space: maintenance contracts, service contracts and warranties. In all three cases, the customer pays for the contract, either as a separate agreement (maintenance and service contracts) or as part of the product purchase price. So, what is the difference between these? Here are the definitions:

- Maintenance contracts cover a predefined set of preventative maintenance activities performed on a predefined schedule. Generally, maintenance contracts don't cover the repair or replacement of the products caused by defects in materials or workmanship, or due to wear and tear.
- Service contracts cover a certain set of clearly defined services that the customer may or may not need. In other words, you are paying for rectifying potential problems that may or may not occur during the contract period. A service contract is like and insurance policy.
- A warranty is an agreement by the warranty provider to restore the product to a specified condition at the warranty provider's expense for any system failures that occur during the warranty period. Extended warranties that are not included with the original product purchase are service contracts.

Chapter 9 discusses contracts for system sustainment in more detail.

4.9 Maintenance Scheduling

The preceding sections in this chapter discussed the conditions under which maintenance should be performed, the optimal interval to perform maintenance, and how many spares are necessary. What we have not addressed is how the availability of maintenance resources influences when maintenance can be performed, i.e., just because CBM or PHM indicate that maintenance needs to be performed, the preventative maintenance interval has been reached, or the system fails, does not necessarily mean that the resources required to do the maintenance are available. The required resources (besides possibly a spare part) include workforce and facilities – these resources are in finite supply, so maintenance may have to wait for the appropriate resources to become available. The other scheduling challenge is the optimal combining of maintenance actions, i.e., it costs time and money to remove a system from service to perform maintenance, so one might like to combine multiple maintenance actions together and perform them at the same time if possible. For some assets, the maintenance resources have to be moved to the asset (e.g., wind turbines), so if the resources are being moved to the wind farm (which could be expensive), you would like to perform as many maintenance actions as practical while the resources are available at the asset.

Maintenance scheduling is the process of making sure that the planned maintenance is performed. Maintenance scheduling brings the required resources together to make sure tasks are completed correctly and on time or in a timely manner. Maintenance scheduling is often confused with maintenance planning. However, the two are different processes; planning deals with what needs to be done and how it will be done, while scheduling handles who does it, when and where it is done. Capacity planning is the determination of the maintenance resources needed to meet a forecasted maintenance load: manpower (number and skills), overtime capacity, contract maintenance, healthy level of backlog, etc.

Maintenance scheduling optimization is a subset of broader project planning models that have been studied extensively with the development of many maintenance optimization models and software packages. We will not address quantitative project planning in this book, however; as a demonstration of the magnitude of maintenance planning problems consider the following. Suppose that you have a small regional airline operating out of one airport. You have a single maintenance facility that can only service one plane at a time. Three of your aircraft require maintenance before they can carry passengers again. What order do you maintain them in? In general, there are $3! = 6$ different sequences (orders) that the maintenance jobs could be performed in. Which sequence is optimal? Let's assume that optimal means least cost and the following simple data where aircrafts A, B and C are all different:

Aircraft	Time to perform maintenance (hours)	Revenue lost per hour out of service
A	4 hours	$5000/hr
B	2 hours	$4000/hr
C	3 hours	$2500/hr

Assuming that the aircraft waiting in the queue for maintenance are assessed the specified revenue loss during their waiting time, the best maintenance order is B, A, C (its total cost is $60,500). What if there are two maintenance facilities that can operate in parallel? Now the best maintenance order is to maintain A and B concurrently and then C in the facility that was used for B (the total cost is $40,500, less the cost of the second facility of course). Obviously, scheduling problems are significantly more complex than this simple example. For example, probably maintenance involves several different activities that may have to be performed in a specific sequence (or may be rearrangeable). The resources required to perform the activities may be independent or overlap (e.g., the workforce and/or other facilities may overlap). In general, if there are n_j jobs requiring m_a independent activities that can be performed in any order, then there are $(m_a!)(n_j!)^m$ possible sequences. So, for the example in this section with three aircraft ($n_j = 3$) and if there were four activities ($m_a = 4$) that could be performed in any order, there would be 31,104 possible unique sequences for maintaining the three aircraft.

4.10 Failure Free and Maintenance Free Operating Periods

A clear understanding of the mechanics of failure, the ruggedness of components, and the operational environment can lead to the derivation of a probability of time in-service before the occurrence of a failure or fault. A Failure Free Operating Period (*FFOP*) is defined as a period of time (or appropriate units) during which no failures, which result in a loss of system functionality, occur [4.15].

During the *FFOP*, faults may exist in the system. A fault is defined as a non-conformance condition that, by itself, may not affect

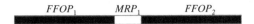

$FFOP_1$ MRP_1 $FFOP_2$

Fig. 4.3.   Failure free operating period (FFOP) and maintenance recovery period (MRP).

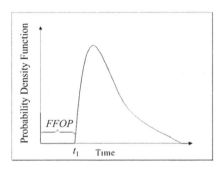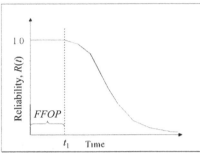

Fig. 4.4.   3-parameter Weibull probability density function (left), reliability for *FFOP* (right).

the required functionality of the system. Over time, these faults may accumulate and lead to a failure hence terminating the *FFOP*. The system may be designed (using redundancy, for example) in such a way that faults do not develop into failures during the operating period. Any faults that do arise are absorbed by the inherent fault tolerance of the system. The faults are rectified during the maintenance activities carried out at the end of the *FFOP* as shown in Figure 4.3.

The system may need to be maintained at some point. This is carried out during a Maintenance Recovery Period (*MRP*). After each designated *FFOP*, there will be an *MRP* that will include all maintenance actions necessary to recover the system to a state, whereby it can be ready to complete the next *FFOP* (Figure 4.3). The length and intensity of the *MRP* will be generally related to the length of the previous and subsequent *FFOP*'s.

Figure 4.4 shows the theoretical probability density function for a system with a *FFOP*. The probability of failure is zero until time t_1.

The reliability of the system is one for time $< t_1$. Failures commence after time t_1 and hence the reliability degrades with increasing time. Therefore, reliability $= 1$ for time $\leq FFOP$, and reliability < 1 for time $> FFOP$. If the time-to-failure is represented as a

3-parameter Weibull distribution, the *FFOP* is the location parameter.

The Maintenance Free Operating Period (*MFOP*) [4.15] is similar in concept to the *FFOP*, but it is a period during which no maintenance is allowed or available.[21] In the case of *FFOP*, maintenance can be done during the *FFOP*. *MFOP* is relevant to systems that cannot be maintained continuously, e.g., offshore wind turbines, oil well drill heads, etc. These systems have to be able to operate maintenance-free for a period of time.

Two quantitative models for predicting the *MFOP* probability have been proposed [4.15]. First, based on Equation (3.23) the probability that a component will survive the duration of the ith *MFOP* cycle, given that it has survived i-1 cycles, is

$$R(t_{MFOP}, (i-1)t_{MFOP}) = \frac{R(it_{MFOP})}{R((i-1)t_{MFOP})} \qquad (4.30)$$

where t_{MFOP} is the duration of the *MFOP*. This equation represents a mission reliability approach to interpreting *MFOP*. Analytical expressions based on Equation (4.30) have been developed and used for system planning and evaluation.

Considering a single part with a Weibull reliability given by Equation (3.21). The probability that a system that has survived to t survives t_{MFOP} longer, and Equation (4.30) becomes,

$$R(t_{MFOP}, t) = \frac{R(t + t_{MFOP})}{R(t)} = e^{-\left(t^{\beta} - (t+t_{MFOP})^{\beta}/\eta^{\beta}\right)} \qquad (4.31)$$

Rearranging Equation (4.31) allows us to solve for t_{MFOP},

$$t_{MFOP} = [t^{\beta} - \eta^{\beta}\ln(R(t + t_{MFOP}, t))]^{1/\beta} - t \qquad (4.32)$$

Equation (4.32) allows us to solve for the t_{MFOP} for a repairable system for a required confidence level when the time-to-failure has a Weibull distribution. For example, if this part has survived 1000

---

[21]Kumar *et al.* [4.15] define *MFOP* as: "a period of operation during which an item will be able to carry out all its assigned missions, without the operator being restricted in any way due to system faults or limitations, with the minimum of maintenance".

hours and you want to know, with a 90% confidence ($R(t_{MFOP}, t) = 0.9$) how long it will last before it needs maintenance, and the part's time-to-failure distribution is characterized by $\eta = 1300$ hours and $\beta = 3.1$, then the *MFOP* duration is $t_{MFOP} = 71.2$ hours.

To understand how *MFOP* can be used for a system of components, consider the simple system consisting of three components in series shown on the left side of Figure 3.7. In this case Equation (4.30) becomes,

$$R(t_{MFOP}, (i-1)t_{MFOP}) = \frac{\prod_{j=1}^{3} R_j(it_{MFOP})}{\prod_{j=1}^{3} R_j((i-1)t_{MFOP})} \qquad (4.33)$$

Assuming that component 1 has a constant failure with $\lambda_1 = 0.002 \, \text{hour}^{-1}$, components 2 and 3 have time-to-failures that have a Weibull distribution with $\eta_2 = 1300$ hours, $\beta_2 = 3.1$, and $\eta_3 = 1000$ hours, $\beta_3 = 2$ ($\gamma = 0$ in both cases), and the $t_{MFOP} = 100$ hours. In this case, the values of R from Equation (4.33) are:

i	Conditional Reliability	Time (hours)
1	0.8103	100
2	0.7924	200
3	0.7729	300
...		
9	0.6264	900

This result means that if the required confidence level is greater than 0.8, maintenance has to be performed after the first cycle (1 *MFOP*, or 100 hours); or if the confidence level required is only 0.6, after 9 *MFOP*s or 900 hours will do.

The treatment in Equations (4.30)–(4.33) implicitly assumes that the maintenance activities are instantaneous. If the maintenance activity requires a non-negligible period of time, an alternating renewal approach is required (Section 4.4.1.3). In this case, a maintenance recovery period governed by the time-to-repair distribution $g(v)$ alternates with the *MFOP*. Assuming that the time-to-failure is characterized by $f(u)$, the probability that the system will have at least t_{MFOP} (that is t_{MFOP} of *MFOP*) during a mission of duration T is given by [4.15]

$$P_1(T) = R(t_{MFOP}) + \int_0^T f(u|t_{MFOP})P_0(T-u)du \qquad (4.34)$$

where

$$P_0(T) = \int_0^T g(v)P_1(T-v)dv \qquad (4.35)$$

where $f(u|t_{MFOP})$ is the value of the PDF at time u given that it has survived up to the t_{MFOP}. Equations (4.34) and (4.35) assume that maintenance is carried out immediately upon failure.

4.11 Summary: Maintenance Policy

As a summary of this chapter, we present a taxonomy of maintenance policies (or strategies) proposed by Cui [4.16]:

- *Time-Dependent (Section 4.5)*: maintenance is performed at specified times determined by system age, or period. This category may use some other degradation-related measure as a surrogate for time (e.g., miles).
- *Degradation-Degree-Dependent (Section 4.6)*: maintenance is only performed when the system's degradation reaches a specified threshold of degradation.
- *Mixed Time and Deterioration Dependent*: maintenance is performed if the system's degradation reaches a specified threshold of degradation OR after as specified period of time, whichever comes first. Other mixed policies are possible.
- *Information-Dependent*: maintenance is carried out based on detected information about the system. This is usually based on a pre-defined stopping rule associated with statistically-dependent systems.
- *Block*: maintenance is only be done when multiple maintenance actions can be grouped together.
- *Priority*: when multiple maintenance actions are needed, this policy specifies the priority order of those maintenance actions.

To these we will add:

- *Corrective (Section 4.4)*: maintenance is only performed when the system fails.

Other maintenance policies exist that are hybrids of those described above. Also, maintenance policies exist that account for imperfect repair. The policy chosen is usually based on the minimization of life-cycle cost, or the maximization of availability or readiness. Maintenance policy is fundamentally determined by a combination of:

- When maintenance is needed
- Where maintenance must be performed
- What has to be maintained
- Availability of the required resources
- What the penalty is for system failure.

References

[4.1] Dhillon, B. S. (1999). *Engineering Maintainability* (Gulf Publishing Company, USA).

[4.2] Economics and the art of maintenance: Repair is as important as innovation. *The Economist*, October 20, 2018.

[4.3] U.S. Department of Defense (1966). *Maintainability Production*, MIL-STD-472.

[4.4] U.S. Department of Defense (2011). *Reliability-Centered Maintenance (RCM) Process*, MIL-STD-3034.

[4.5] Palm, C. (1938). Analysis of the Erlang traffic formulae for busy-signal arrangements. *Ericsson Technics*, 4, pp. 39–58.

[4.6] Cox, D. R. (1962). *Renewal Theory* (Methuen & Co, UK), p. 40.

[4.7] Smith, W. L. (1954). Asymptotic renewal theorems, *Proceedings of the Royal Society*, 64, pp. 9–48.

[4.8] Blischke, W. R. and Murthy, D. N. P. (1994). *Warranty Cost Analysis* (Marcel Dekker, New York).

[4.9] Rausand, M. and Høyland, A. (2003). *System Reliability Theory: Models, Statistical Methods, and Applications*, 2nd Ed. (John Wiley & Sons, USA), pp. 275–277.

[4.10] Elsyed, E. A. (1996). *Reliability Engineering* (Addison Wesley Longman, USA).

[4.11] Williams, J. H., Davies, A., and Drake, P. R. (eds.) (1994). *Condition-Based Maintenance and Machine Diagnostics* (Chapman & Hall, UK).

[4.12] Feldman, K., Jazouli, T., and Sandborn, P. A. (2009). A methodology for determining the return on investment associated with prognostics and health management, *IEEE Transactions on Reliability*, 58(2), pp. 305–316.

[4.13] Moubra, J. (1997). *Reliability-Centered Maintenance* (Industrial Press, USA), p. 496.

[4.14] *SAE JA1011, Evaluation Criteria for Reliability-Centered Maintenance (RCM) Processes*, Society of Automotive Engineers, 1 August 1998.

[4.15] Kumar, D. U., Knezevic, J., and Crocker, J. (1999). Maintenance free operating period – An alternative measure to MTBF and failure rate for specifying reliability? *Reliability Engineering & System Safety*, 64, pp. 127–131.

[4.16] Cui, L. (2008). Maintenance models and optimization, In: *Handbook of Performability Engineering*, K. B. Misra (ed.) (Springer, UK).

Problems

4.1 A system consists of three replaceable components described by the following data:

Component	$MTBF$ (hours)	$MTTR$ (hours)
A	500	0.5
B	600	1.3
C	1000	1.0

Assume that the $MTTR$ and $MTBF$ are represented by exponential distributions. What is the $MTTR$ of this system?

4.2 Suppose that a part to be spared is used in three different systems. The part has a system-specific constant failure rate and a system-specific usage profile as shown in the table on the next page. If spare parts are procured every 200 days, how many spares should be carried in inventory to ensure a 90% protection level?

System	n	λ (failures/1000 operating hours)	Operating hours per day
A	26	0.11	12
B	23	0.065	18
C	43	0.16	20

(a) Assume that all the systems draw from the same pool of spare parts.

(b) Assume that each system draws from its own, unique pool of spare parts that the other systems cannot draw from.

4.3 If the MTBF of a voting machine is 163 hours, and assuming failed voting machines are fixed to good-as-new condition instantaneously when they fail, how many total voting machine failures will there be in New York City in one 15-hour election day? Assume New York City simultaneously supports a total of 7531 voting machines on an election day.

4.4 A constant-interval replacement reliability problem presented in the text in the 3rd paragraph of Section 4.5. Solve this problem using Equation (4.26).

4.5 Prove that $t_p = \infty$ is the optimum preventative maintenance interval for the bus problem in Section 4.4.1.2 using Equation (4.27).

4.6 Assume a solar-powered remote railroad crossing signal has 10 solar cells connected in parallel. The signal can operate correctly with a minimum of 3 cells working. Assume that the reliability of the cells is independent and each cell has a constant failure rate of 1 failure/year. What is the probability that the signal can operate without failure for a minimum of 2 years?

4.7 In the simple maintenance sequencing problem in Section 4.9, what is the worst order to perform maintenance in?

4.8 Derive the t_{MFOP} for a constant failure rate (i.e., derive Equation (4.32) for a constant failure rate).

4.9 In the *MFOP* example in Section 4.10, what confidence corresponds to five *MFOP* cycles?

Chapter 5

Availability and Readiness

Availability is the ability of a service or a system to be functional when it is requested for use or operation. The concept of availability accounts for both the frequency of failure (reliability) and the ability to restore the service or system to operation after a failure (maintainability). The maintenance ramifications generally translate into how quickly the system can be repaired upon failure and are usually driven by logistics management. The concept of availability only applies to "repairable" or "restorable" systems that are either externally maintained or self-maintained.

Availability has been an important design parameter for the critical-system communities for many years, but recently it is beginning to be recognized, quantified, and studied for other types of systems. Many real-world systems are significantly impacted by availability. A decrease in availability of an ATM machine causes inconvenience to customers; poor availability of wind farms can make them unprofitable; the unavailability of a point-of-sale system to retail outlets can generate a huge financial loss; the unavailability of a medical device or of hospital equipment can result in loss of life. For web-based business services, the availability of a web site and the data to support depends on the reliability and maintainability of servers. In these example systems, insuring the availability of the system becomes the primary interest and the owners of the systems are often willing to pay a premium (purchase price and/or support) for higher availability.

5.1   Time-based Availability Measures

Reliability is the probability that an item will not fail; maintainability is the probability that a failed item can be successfully restored to operation. Availability is the probability that an item will be able to function (i.e., not be failed or undergoing repair) when called upon to do so over a specific period of time under stated conditions. Measuring availability provides information about how efficiently a system is supported.

In general, time-based availability is computed as the ratio of the uptime to the sum of the uptime and downtime:

$$A = \frac{uptime}{uptime + downtime} \tag{5.1}$$

where *uptime* is the total expected operational time during which the system is "up" and running and able to perform the tasks that are expected from it; *downtime* is the expected period that the system is "down" and not operating when requested due to repair, replacement, waiting for spares, or any other logistics or administrative delays. The sum of the accumulated uptime and downtime represents the total operation time for the system.[1]

5.1.1   *Time-interval-based availability measures*

If the primary concern is a time interval, then we consider instantaneous, average, and steady-state availability.

Instantaneous (also called point or pointwise) availability is the probability that an item will be able to perform its required function at the instant it is required. Instantaneous availability at time t is given by:

$$A(t) = R(t) + \int_0^t R(t - \tau)m(\tau)d\tau \tag{5.2}$$

where

---

[1]Equation (5.1) implicitly assumes that uptime is equal to operational time, whereas in reality, not all of the uptime is actually operational time; some of it corresponds to time the system spends in standby mode waiting to operate.

$R(t)$ is the reliability at time t, (the probability that the item functioned without failure from time 0 to t),

$R(t - \tau)$ is the probability that the item functioned without failure since the last repair time τ, and

$m(\tau)$ is the renewal density function.

Equation (5.2) represents a sum of probabilities. The first term is the probability of no failure occurring from time 0 to t, the second term (the integral) is the probability of no failure since the last repair time (τ).

A renewal function, $M(t)$, (see Section 4.4.1.2) is the expected number of failures in a population. The renewal density function is the mean number of renewals expected in a narrow interval of time near t: $m(t) = dM(t)/dt$. In general, the renewal density function can be found by taking the derivative of Equation (4.15),

$$\hat{m}(s) = \frac{\hat{w}(s)\hat{g}(s)}{1 - \hat{w}(s)\hat{g}(s)} \tag{5.3}$$

where $\hat{m}(s)$ is the Laplace transform of $m(t)$, and $\hat{w}(s)$ and $\hat{g}(s)$ are the Laplace transforms of the time-to-failure distribution and time-to-repair distributions, respectively.[2] Using Equation (5.3) and Equation (5.2), the Laplace transform of the availability becomes

$$\hat{A}(s) = \frac{1 - \hat{w}(s)}{s(1 - \hat{w}(s)\hat{g}(s))} \tag{5.4}$$

Instantaneous availability is a useful measure for systems that are idle for periods of time and then are required to perform at a random time, such as a defibrillation unit in a hospital or a torpedo in a submarine.

The average (also called mean, average uptime, or interval) availability is given by

$$\overline{A(t)} = \frac{1}{t} \int_0^t A(\tau)d\tau \tag{5.5}$$

---

[2]Note, we are using traditional notation here, with w representing the time-to-failure distribution. $\hat{w}(s)$ is the same as $\hat{f}(s)$ in Chapter 4.

The average availability in Equation (5.5) is the proportion of time in the interval $[0, t)$ that the system is available. Average availability is used for systems whose usage is defined by a duty cycle, like a commercial airliner or construction equipment at a job site.

The steady-state (or limiting) availability is given by

$$A(\infty) = \lim_{t \to \infty} A(t) \tag{5.6}$$

where $A(t)$ is the instantaneous availability. Equation (5.6) is only valid if the limit exists. Steady-state availability is often applied to systems that operate continuously – for example, an air traffic control radar system or a computer server. As a general rule, the instantaneous availability will start approaching the steady-state availability after a time period of approximately four times the average time to failure of the system.

5.1.2 *Downtime-based availability measures*

Availability measures that focus on the various mechanisms that contribute to downtime include inherent availability, achieved availability, and operational availability. The relevant time measures are summarized in Table 4.2. Availability measures in this category are distinguished based on the activities included in the downtime and have the general form shown in Equation (5.1). All of these availability measures assume a steady-state condition.

Inherent (or intrinsic) availability is defined as

$$A_i = \frac{MTBF}{MTBF + MTTR} \tag{5.7}$$

where $MTBF$ is the mean time between failures and $MTTR$ is the mean time to repair (or mean corrective maintenance time). Inherent availability only includes downtime due to corrective maintenance actions (excluding preventative maintenance, logistics, and administrative downtimes). Inherent availability is used to model an ideal support environment.

Achieved availability is given by

$$A_a = \frac{MTBM}{MTBM + \overline{M}} \tag{5.8}$$

where $MTBM$ is the mean time between maintenance activities and $\overline{M}$ is the mean active maintenance time. Achieved availability is also used to model an ideal support environment.

Operational availability is the availability that the customer actually experiences in a real operational environment:

$$A_o = \frac{MTBM}{MTBM + MDT} \tag{5.9}$$

The denominator of Equation (5.9) is the overall operational time period. Operational availability is used to model an actual (non-ideal) support environment.

A common availability metric used in inventory analysis is supply availability, which is defined as

$$A_s = \frac{MTBM}{MTBM + MSD} \tag{5.10}$$

The denominator of Equation (5.10) specifically excludes the time associated with diagnosing or making a repair – that is, it is independent of the maintenance policy and only depends on the sparing policy for stocking spares. Supply availability is specifically discussed in Section 5.2.1.

As an example of availability estimation using downtime-based availability measures, consider an electronic system with the following characteristics ("op hours" = operational hours):

- Operational cycle = 2000 op hours/year
- Support life = 5 years
- Failures that require corrective maintenance = 2/year
- Repair time per failure = 40 op hours
- Preventative maintenance activities = 1/year
- Preventative maintenance time per preventative maintenance action = 8 op hours
- Average wait time for repair materials for corrective maintenance = 10 op hours

From the given information, $MTTR = 40$ op hours, $MTPM = 8$ op hours, $LDT = 10$ op hours, and the following quantities can be calculated:

$$\text{Total number of maintenance actions}$$
$$= (2)(5) + (1)(5) = 15 \tag{5.11a}$$

$$\overline{M} = \frac{(40)(2)(5) + (8)(1)(5)}{15} = 29.333 \text{ op hours} \tag{5.11b}$$

$$MDT = \frac{(40 + 10)(2)(5) + (8)(1)(5)}{15} = 36 \text{ op hours} \tag{5.11c}$$

$$MTBF = \frac{(5)(2000)}{(2)(5)} = 1000 \text{ op hours} \tag{5.11d}$$

$$\text{Total operational cycle} = (5)(2000) = 10{,}000 \text{ op hours} \tag{5.11e}$$

$$\text{Total downtime} = (15)(36) = 540 \text{ op hours} \tag{5.11f}$$

$$\text{Total uptime} = 10{,}000 - 540 = 9460 \text{ op hours} \tag{5.11g}$$

$$MTBM = \frac{9460}{15} = 630.667 \text{ op hours} \tag{5.11h}$$

Using the quantities in Equation (5.11), we can calculate the availabilities as:

$$A_i = \frac{1000}{1000 + 40} = 0.9615 \tag{5.12a}$$

$$A_a = \frac{630.667}{630.667 + 29.333} = 0.9556 \tag{5.12b}$$

$$A_o = \frac{630.667}{630.667 + 36} = 0.9460 \text{ or } A_o = \frac{9460}{10{,}000} = 0.9460 \tag{5.12c}$$

Notice that the same operational availability is computed two different ways in Equation (5.12c).

5.1.3 *Application-specific availability measures*

Several additional specialized types of time-based availability also exist. These availability measures represent the availability for specific applications:

- *Mission availability*: The probability that each failure occurring during a mission of a specific total operating time can be repaired in a time that is less than or equal to some specified time length. Mission availability is applicable to situations when only a finite amount of repair time is acceptable.

- *Work-mission availability*: The probability that the sum of all the repair times for all the failures occurring in a mission of a specified total operating time is less than or equal to some specified time length.
- *Joint availability*: The probability of the system operating at two distinct times during a mission.
- *Random-request availability*: Incorporates the performance of several tasks arriving randomly during the fixed mission period. Random-request availability includes both the system state and random task arrival rates.
- *Computation availability*: The mean performance level at a given time, which is the weighted sum of state probabilities.

5.1.4 *System availability*

Since availability is a probability, it is accumulated similarly to reliability. Series systems, Figure 5.1, are systems that only operate satisfactorily if all the components in the system are operating. In a series system, the system availability is simply the product of the component availabilities,

$$A = A_1 A_2 \ldots A_n \qquad (5.13)$$

Equation (5.13) indicates that the combined availability of multiple components in series is always lower than the availability of the individual components.

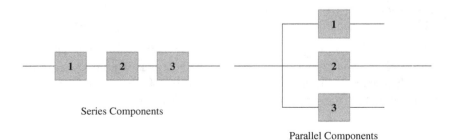

Fig. 5.1.   Series and parallel components.

If the components are in parallel, Figure 5.1, the system availability is,

$$A = 1 - [1 - A_1][1 - A_2] \cdots [1 - A_n] \qquad (5.14)$$

Parallel operation provides a powerful mechanism for making a highly available system from low availability parts. For this reason, many mission-critical systems are designed with redundant subsystems. Other sorts of systems that are composed of both series and parallel components are aggregates of Equations (5.13) and (5.14).

5.1.5 *Partial system availability*

For some types systems, availability is not black or white, i.e., the system isn't available or not available, rather it is partially available. Consider for example a cluster of servers, a call processing system, a wind farm of turbines, or a fleet of buses. These sorts of systems are designed so that not all the members of the population have to be operational all the time in order the system-of-systems to accomplish its mission.

For example, consider a cluster of 3 servers where a minimum of 2 servers need to be available at any time for the cluster to have satisfactory performance. In this case, if the availability of each server is given by A_{sys} the probability that all 3 servers are available at the same time is given by A_{sys}^3. The probability that 2 servers are available and one specific server (server #1 for example) is not available is given by, $A_{sys}^2(1 - A_{sys})$, but there are 3 possible combinations of 2 servers available and 1 not available So, if the servers are indistinguishable, the probability that 2 servers are available and exactly 1 (any one of the 3 servers) is unavailable is given by, $3A_{sys}^2(1 - A_{sys})$. If you add to this the case where all 3 are available, you get, $A_{sys}^3 + 3A_{sys}^2(1 - A_{sys})$, which is the availability of server cluster if at least 2 of the 3 servers have to be available at any given time. In general, the availability of a system of N things where maximum of L that can be unavailable ($L \leq N$) is given by,

$$A = \sum_{i=0}^{L} \frac{N!}{i!(N-i)!} A_{sys}^{N-i}(1 - A_{sys})^i \qquad (5.15)$$

Each term in Equation (5.15) is the probability that exactly i things out of N things are unavailable. The sum provides the probability that L or fewer things out of N things are unavailable.

For the server example given above, $N = 3$ servers in the cluster, a minimum of 2 servers must be available (therefore, $L = N - 2 = 1$ server can be unavailable), if the servers are indistinguishable and their individual availabilities are given by $A_{sys} = 0.9$ (and the causes of server unavailability are independent), then the availability of this system of servers is,

$$A = (1)(0.9^3)(1 - 0.9)^0 + (1)(0.9^3)(1 - 0.9)^1 = 0.9720 \qquad (5.16)$$

In Equation (5.16) the first term is the probability that all 3 servers are available, the second term is the probability that exactly 2 out of 3 are available. So, if the cluster is considered "up" if 2 or more servers are available, the probability of achieving this with a 3-server cluster is 0.9720, or the cluster has a 0.9720 availability.

Here is slightly different example: suppose you have a fleet of 20 helicopters and the availability of each individual helicopter is 0.95, what is the probability that at least 18 helicopters in the fleet are available at an instant in time? In this case $N = 20$, $A_{sys} = 0.95$, and $L = 20 - 18 = 2$; the resulting system availability given by Equation (5.15) is 0.9245. Note, if $L = 0$ then $A = A_{sys}^{20} = 0.3585$; if $L = 20$ then $A = 1$.

5.1.6 *Readiness*

Availability and readiness relate a system's operating time between failures to some other longer period of time; the difference between the two concepts is what is included in the longer period of time. "Availability" is defined in terms of operating time and downtime, where downtime includes active repair time, administrative time, and logistic time. Alternatively, operational readiness is defined in terms of a longer time period that is all calendar time.

Consider the bus fleet discussed in Section 4.6 (with and without PHM included). In this case the availability/readiness of the bus fleet may also be a relevant issue; a simple availability calculation for this case, using Equation (5.1) for availability assuming uptime + downtime = all time (making this actually a readiness calculation), the availabilities with and without PHM become,

$$A_{no\,PHM} = \frac{(24)(7)(365) - (3)(5)(24)}{(24)(7)(365)} = 0.9941$$

$$A_{PHM} = \frac{(24)(7)(365) - [(1)(5)(24) + (2)(0.5)(24)]}{(24)(7)(365)} = 0.9977$$

where the data in Section 4.6 has been used for the calculations above.

Readiness can be quantitatively expressed as [5.1]:

$$A_R = A_{sys} + (1 - A_{sys})\mathrm{Pr}(t_m < t_d) \qquad (5.17)$$

where t_m is the maintenance time (all the downtimes included in the calculation of availability), t_d is called the relaxation time (the duration from the end of maintenance to the start of the next mission), and $\mathrm{Pr}(t_m < t_d)$ is the probability that t_d is greater than t_m. If $t_d = 0$ then $A_R = A_{sys}$. In general, the calculation of $\mathrm{Pr}(t_m < t_d)$ depends on the distributions associated with t_m and t_d. If both t_m and t_d are exponentially distributed,

$$f(t_m) = \mu_{rr}e^{-\mu_{rr}t_m} \text{ and } f(t_d) = \lambda e^{-\lambda t_d} \qquad (5.18)$$

the $\mathrm{Pr}(t_m < t_d)$ becomes,

$$\mathrm{Pr}(t_m < t_d) = \frac{\mu_{rr}}{\mu_{rr} + \lambda} \qquad (5.19)$$

If the $MTBM = 1000$ hours, $MDT = 17.3$ hours, $MTTR = 2$ hours ($\mu_{rr} = 1/MTTR$), and the mean of t_d distribution (λ) $= 0.0833$ hours^{-1}, then the operational availability from Equation (5.9) is $A_o = 0.9830$ and from Equation (5.17), $A_R = 0.9976$.

Equation (5.17) can also be written in the following form [5.1],

$$A_R = R(t) + F_D(1 - R(t))M_a(t_d) \qquad (5.20)$$

where F_D is the fault detection rate, $M_a(t_d)$ is the maintainability given by Equation (4.2), $R(t)$ is reliability. Equation (5.20) can be used to determine the required fault detection rates to achieve a specific level of readiness.

5.1.7 *Markov availability models*

Markovian approaches to the formulation of availability models are also widely used. The simplest Markov model is the Markov chain, which models the state of a system with a random variable that changes over time. In this context, the Markov property suggests that the distribution for the variable depends only on the distribution of the variable in the previous state.[3]

Consider a simple two-state model where $X(T)$ represents the status of the system (S) at time T. Assume that $X(T) = 0$ means the system is down (not available) at time T, and $X(T) = 1$ means the system is up (available) at time T. The state transition diagram for our system S is shown in Figure 5.2.

The state transition probabilities in Figure 5.2 are given by p_{ij}, which is the probability that the state is j at T, given that it was i at time $T - 1$. The state transition probabilities in Figure 5.2 are given by

$$p_{01} = \Pr[X(T) = 1 | X(T - 1) = 0] = q$$
$$p_{10} = \Pr[X(T) = 0 | X(T - 1) = 1] = p$$
$$p_{00} = \Pr[X(T) = 0 | X(T - 1) = 0] = 1 - q$$
$$p_{11} = \Pr[X(T) = 1 | X(T - 1) = 1] = 1 - p$$

where p and q represent Markov chain one-stage transition probabilities and $p_{00} + p_{01} = 1$ and $p_{10} + p_{11} = 1$, since there are only two states the system can be in (up or down).

Markov chains can be represented using a state transition probability matrix like the one constructed in Figure 5.3.

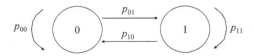

Fig. 5.2.   State transition diagram for system S.

---

[3] Markov processes are "memoryless", i.e., the probability distribution of the next state depends only on the current state and not on the sequence of events that preceded it.

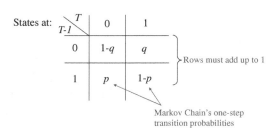

Fig. 5.3.   State transition matrix construction.

The state transition probability matrix for our simple system represents the probabilities of moving from one state to any other state, and is given by

$$\mathbb{A}_t = \begin{bmatrix} 1-q & q \\ p & 1-p \end{bmatrix} \tag{5.21}$$

If we need to determine the probabilities of moving from one state to another state in two steps, all we have to do is raise Equation (5.21) to the second power:

$$\begin{bmatrix} 1-q & q \\ p & 1-p \end{bmatrix}^2 = \begin{bmatrix} 1-q & q \\ p & 1-p \end{bmatrix}\begin{bmatrix} 1-q & q \\ p & 1-p \end{bmatrix}$$

$$= \begin{bmatrix} (1-q)^2 + qp & (1-q)q + q(1-p) \\ p(1-q) + (1-p)p & pq + (1-p)^2 \end{bmatrix} = \begin{bmatrix} p_{00}^2 & p_{01}^2 \\ p_{10}^2 & p_{11}^2 \end{bmatrix} \tag{5.22}$$

Note that a matrix multiplication is used in Equation (5.22). For example, the probability p_{10}^2 in Equation (5.22) represents the probability that system S is down after operating for $T = 2$ time steps if it was initially up (in state 1). Note that the rows of the state transition probability matrix in Equation (5.22) still add up to one.

For large n, the state transition matrix has quasi-identical rows and the results are interpreted as "long run averages" or "limiting probabilities" of S being in the state corresponding to column i:

$$\begin{bmatrix} 1-q & q \\ p & 1-p \end{bmatrix}^n = \frac{1}{p+q}\begin{bmatrix} p & q \\ p & q \end{bmatrix} + \frac{(1-p-q)^n}{p+q}\begin{bmatrix} q & -q \\ -p & p \end{bmatrix} \tag{5.23}$$

In the limit as n approaches infinity,

$$\lim_{n\to\infty} \begin{bmatrix} 1-q & q \\ p & 1-p \end{bmatrix}^n = \frac{1}{p+q} \begin{bmatrix} p & q \\ p & q \end{bmatrix} \tag{5.24}$$

For the example, if $MTBF = 600$ and an $MTTR = 34$,

$p = p_{10} = 1/600 = 0.00167$ (probability of failing is $1/MTBF$)

$q = p_{01} = 1/34 = 0.0294$ (probability of being repaired is $1/MTTR$)

The transition probabilities in this case are given by

$$p_{11}^n = p_{01}^n = \frac{q}{p+q} = 0.9464$$

$$p_{00}^n = p_{10}^n = \frac{p}{p+q} = 0.0536$$

Thus p_{11}^n and p_{00}^n are state occupancy rates, which can also be interpreted as the fraction of time that the system will spend in the "up" and "down" states respectively – that is, the expected availability and unavailability of the system. In this case the inherent availability is p_{11}^n, note, $600/(600 + 34) = 0.9464$.

5.2 Non-time-based Availability

One way to view availability is operational (time based), while an alternative view is through the lens of demand. Viewing availability as the ability to support a system when the demand for the system arrives, leads us to the consideration of availability as an inventory problem. MDT discussed in Sections 4.3 and 5.1.2 depends on both the time to perform a repair and the availability of spare parts (the spare part stocking or inventory level).

Sections 5.2.1 and 5.2.2 address the challenge of determining the minimum number of spares necessary to meet an availability requirement. Section 5.2.3 is also an inventory view of availability, but one in which the inventory is the fielded systems (not spare parts); and Section 5.2.4 is a discussion of energy availability used for energy generation sources.

5.2.1 *Backorders and supply availability*

A backorder is an unfulfilled demand due to the lack of spares. The probability of a population of n items having exactly x failures in time t is given by the $i = x$ term in Equation (4.13),

$$\Pr(x) = \frac{(n\lambda t)^x}{x!} e^{-\lambda t} \tag{5.25}$$

If k spares exist for this population, then the probability of needing exactly $k + m_b$ spares resulting in a backorder of m_b is given by

$$\Pr(k + m_b) = \frac{(n\lambda t)^{k+m_b} e^{-n\lambda t}}{(k + m_b)!} \tag{5.26}$$

The expected number of backorders for the population of items with k available spares is,

$$\text{EBO}(k) = \sum_{x=k+1}^{\infty} (x - k)\Pr(x) \tag{5.27}$$

where $\Pr(x)$ is given by Equation (5.26). Each of the terms in the sum in Equation (5.27) is the probability of needing $1, 2, 3, \ldots, \infty$ more spares than you have multiplied by that number of spares.[4]

As an example, if there are $n\lambda t = 20$ demands for spares and you have $k = 10$ spares, then the expected number of backorders from Equation (5.27) is $\text{EBO}(10) = 10.01$. EBO is discussed again in Section 6.1.1.

Now we can relate the expected number of backorders to the supply availability (A_s) using [5.2]:

$$A_s = \prod_{i=1}^{l} \left[1 - \frac{\text{EBO}_i(k_i)}{N Z_i} \right]^{Z_i} \tag{5.28}$$

where

l is the number of unique repairable items in the system,
N is the number of instances of the system,

---

[4]A computational simpler (but identical) form of Equation (5.27) is,

$$\text{EBO}(k) = \lambda t - k + \sum_{x=0}^{k} (x - k)\Pr(x)$$

Z_i is the number of instances of item i in each system, and
$EBO_i(k_i)$ is the expected number of backorders for the ith item
if k_i spares exist (this is the total expected backorders
for all instances of the ith item in N systems).

In Equation (5.28), the product NZ_i is n, which is the number
of sockets for the ith item in the N systems (i.e., the number of
places that the ith repairable item occupies in the system). The ratio
$EBO_i(k_i)/NZ_i$ is the probability of an unfulfilled spare demand for
the entire population of the ith item. Then, $1\text{-}EBO_i(k_i)/NZ_i$ is the
probability that there are no unfulfilled spare demands in the entire
population of the ith item. Raising this quantity to the power Z_i
gives the probability of no unfulfilled spare demands for the ith item
in one instance of the system. That is, the system is assumed to be
available only if there are no unfulfilled spares in the Z_i items of the
ith type in the system. The product in Equation (5.28) assumes that
all l unique repairable items that make up one instance of the system
have to function for the system to be available, so A_s represents the
supply available for the system.

Equation (5.28) assumes that all the i items have independent
failures and that the N systems are independent as well. Also, there
is no cannibalization (i.e., no failed systems are robbed for parts to
fix other systems). Equation (5.28) only applies if $EBO_i(k_i) \leq NZ_i$
for all i.

Consider an example, if there are 1000 systems, each containing 2
unique repairable items (one instance of item 1 and three instances
of item 2), that must be spared for 60 days, and item 1 experiences
twenty demands during the time period and has ten spares, while
item 2 experiences seventeen demands during the time period and
has twelve spares, what is the supply availability for each system in
the fleet? In this case,

$$N = 1000 \qquad Z_1 = 1 \qquad Z_2 = 3$$
$$l = 2 \qquad n\lambda_1 t = 20 \qquad n\lambda_2 t = 17$$
$$k_1 = 10 \qquad k_2 = 12$$

From Equation (5.27) $EBO_1(10) = 10.1$ and $EBO_2(12) = 5.18$. Using
Equation (5.28), the supply availability is given by

$$A_s = \left[1 - \frac{10.1}{(1000)(1)}\right]^1 \left[1 - \frac{5.18}{(1000)(3)}\right]^3 = 0.9848.$$

5.2.2   *Erlang-B*

One way to relate availability to spares is to use the Erlang-B (also known as the Erlang loss formula) [5.3]. This formula was originally developed for planning telephone networks, and it is used to estimate the stock-out probability for a single-echelon repairable inventory[5]:

$$1 - \overline{A} = \frac{a_r^k / k!}{\sum_{x=0}^{k} (a_r^x / x!)} \tag{5.29}$$

where

$\overline{A}$ is the steady-state availability ($1 - \overline{A}$ is the unavailability)
a_r is the number of units under repair
k is the number of spares

In Equation (5.29) $1 - \overline{A}$ is the stock-out probability.[6] The number of units under repair can be computed from

$$a_r = N F_t (MTTR) \tag{5.30}$$

where

N is the number of fielded units,
F_t is the failures that need to be repaired per unit per unit time, and
$MTTR$ is the mean repair time (mean time to repair one unit).

The product NF_t is the arrival rate, or the number of repair requests per unit time. Equation (5.29) assumes that a_r follows a

---

[5]Single-echelon repairable inventory means that the members of the lowest echelon are responsible for their own stocking policies, independent of each other and independent of a centralized depot. Single-echelon means we are basically dealing with a single inventory (or stocking point) of spares. Multi-echelon inventory considers multiple stocking points coupled together (multiple distribution centers and layers), e.g., a centralized depot that provides common stock to multiple lower stocking points. See Chapter 6 for more discussion of inventory echelons.

[6]For telephone networks, $1 - \overline{A}$ is called the blocking probability, the probability of all the servers being busy and a call being blocked (lost). a_r is the traffic offered to the group measured in Erlangs, and k is the number of trunks in the full availability group. Equation (5.29) is used to determine the number of trunks (k) needed to deliver a specified service level ($1 - \overline{A}$), given the traffic intensity (a_r). In general, this formula describes a probability in a queuing system.

Poisson process and is derived assuming that the number of spares (k) is equal to the number of fielded systems requesting a spare (see [5.4]).

As an example of the usage of Equation (5.27), consider a population of 3000 systems where each system has a failure rate of $\lambda = 7 \times 10^{-6}$ failures/hour; 50% of the failures require repair (the other 50% are assumed to either result in system retirement or are resolved with permanent spares taken from another source outside the scope of this problem); the mean repair time is 72 hours. We want a 99.9% availability. How many spares are needed?

$F_t = 0.5\lambda = 3.5 \times 10^{-6}$ failures per unit per hour

$a_r = (3000)(3.5 \times 10^{-6})(72) = 0.756$ the number of units under repair at any one time (this unit of measure is referred to as an Erlang)

$1 - \overline{A} = 0.001$

Applying Equation (5.29), we find that when $k = 5$, $1 - \overline{A} = 0.00097$ (which is less than 0.001), 5 or more spares are needed.

5.2.3 *Materiel availability*

Materiel or matériel is equipment, apparatus, and supplies used by an organization or institution, often specifically associated with a military application. Materiel availability is the fraction of the total inventory of a system that is operationally capable (ready for tasking) for performing a required mission at a specific point in time governed by the condition of the materiel. The key word in this definition is "inventory". If I have an inventory of 10 helicopters and 8 are currently operational and ready for use, then my materiel availability is 0.8 or 80%.

The point or instantaneous materiel availability is expressed as the fraction of end items that are operational, which can be calculated using either of the following relations,

$$A_m = \frac{\text{Number of Operational End Items}}{\text{Total Population of End Items Fielded (in Inventory)}} \quad (5.31a)$$

$$A_m = \frac{\text{Active Inventory}}{\text{Active Inventory} + \text{Inactive Inventory}} \quad (5.31b)$$

Materiel availability is distinguished from time-based availability measures by the fact that it depends on the total population of systems (end items) fielded (in inventory) and it considers the total life cycle of the system (end item).[7]

The materiel availability can be calculated using Equation (5.1), however, the uptime and downtime have different definitions and the materiel availability is not interchangeable with the operational availability. The materiel availability must apply to the entire fielded inventory of systems, apply to the entire life cycle of the system, and incorporate all categories of downtime. Operational availability always applies to a limited number of systems and frequently incorporates only unscheduled maintenance downtimes. A_m is a function of A_o and other factors that do not impact A_o, including technology insertion. While A_o is an operational measure, A_m is a programmatic measure that spans a larger timeframe, additional sources of downtime, and additional sources of unscheduled maintenance.

5.2.4 *Energy-based availability*

Specific applications have discovered that time-based availability measures do not always adequately represent their needs. For example, in the renewable energy generation domain, time-based availability does not account for the fact that the system is not producing efficiently all the time, i.e., just because the system is operating does not mean it is operating efficiently. Conversely, just because the system is not operating does not mean that energy could be produced if it was operational. For example, for a wind farm 3% unavailability when there isn't much wind could represent very little energy loss. While the same unavailability could represent a significant loss during high wind periods [5.5].

While time-based availability[8] is used for renewable energy applications, energy-based availability measures like the following are also

---

[7]Since the definition of materiel availability mandates that it consider the entire fielded population of systems and the entire system life cycle, technically it is impossible to measure until after a system has completed its entire field life.

[8]The term "availability factor" is often used to mean operational availability in power plants.

widely used,

$$A_E = \frac{\text{Available Energy}}{\text{Available Energy} + \text{Energy Lost}} \qquad (5.32a)$$

$$A_E = \frac{E_{\text{real}}}{E_{\text{theoretical}}} \qquad (5.32b)$$

5.3 System Effectiveness

Effectiveness is a figure of merit for judging the opportunity for producing the intended results (i.e., it is a measure of value received). Effectiveness is nominally a probability and is often calculated as a product or two or more of the following values: availability, reliability, maintainability, capability, dependability, performance (where all of these are assumed to be probabilities). In some disciplines, effectiveness is used in practice as a "benchmark".

There is no single definition of system effectiveness for critical systems, but it is a function of availability (reliability and maintainability), and the system's ability to meet the performance/capability requirements bestowed upon it.[9] For example, a car could have great availability (always available when you need it), but poor effectiveness (it only seats two people and you have a family of four).

System effectiveness is the probability that a system will successfully meet all design objectives when called upon to do so. Availability and other measures implicitly assume that a system meets its design objectives if it is operationally available, but this may not be the case. Relevant system design objectives (or performance parameters) vary depending on the type of system and its application. Example parameters might include power generation capacity of a power plant, the range or speed of an aircraft, the fuel efficiency of a car, the destructive yield of a weapon, or the accuracy of a sensor.

---

[9]System effectiveness (in the context of critical systems) first appeared in the mid-1960s in U.S. Department of Defense documents [5.6]. System effectiveness is also a phrase often used to measure organizational effectiveness, which is outside the scope of this treatment.

System effectiveness can be quantitatively defined in several ways. One possible measure is:

$$SE = \frac{N_{AC}}{N_T} \qquad (5.33)$$

where N_{AC} is the number of systems that have met their design objectives and N_T is the total number of systems at the start of the mission.[10] How one determines if a system has met its design objectives is obviously highly application specific. Equation (5.33) can be used, for example, to determine the number of systems that you must have (N_T) if the SE can be quantified. SE could be determined using,

$$SE = AD_A \qquad (5.34)$$

where A is the availability and D_A is the design adequacy. If, for example, $A = 0.98$ and $D_A = 0.84$, then $SE = 0.823$. If we need 100 systems that meet design objectives, then $N_T = 122$ systems have to start the mission. The particular availability measure used in Equation (5.34) depends on how the "mission" is defined.

5.4   Mapping Availability to Cost

Unquestionably there is a correlation between the availability of a system and the system's life-cycle costs. If the system is managed under an availability contract (see Chapter 9), then the mapping may be straightforward, e.g., each hour of unavailability results in a penalty with a specified financial value, or periods of time when the availability drops below a contractually specified availability threshold result in reduction in money collected by the sustainer (again a penalty).

---

[10]There are alternative definitions of system effectiveness exist. One alternative [5.7] is $SE = $ (Effectiveness)/(Life-cycle cost), where Effectiveness is a function of availability, reliability, maintainability and capability, although it is not clear what this function should be. Overall Equipment Effectiveness (OEE) also exists in the context of production equipment. OEE is often calculated as the product of availability, performance and quality.

An alternative way to assess the life-cycle cost impact of availability is to consider availability's impact on the quantity of an asset that must be owned and sustained. Consider the example at the end of Section 5.1.5: you have a fleet of 20 helicopters and the availability of each individual helicopter is 0.95, what is the probability that at least 18 helicopters in the fleet are available at an instant in time? In this case $N = 20$, $A_{sys} = 0.95$, and $L = 20 - 18 = 2$; the resulting system availability given by Equation (5.15) is 0.9245. To establish the monetary value of availability (in this case), we could change the number of helicopters in the fleet to 21 and solve Equation (5.15) for the value of A_{sys} that results in the same overall system availability (0.924516). Setting $N = 21$, $L = 21 - 18 = 3$, and $A = 0.924516$ in Equation (5.15), we get $A_{sys} = 0.922217$. So, 1 helicopter is worth $0.95 - 0.922217 = 0.027783$ in availability (this is the "marginal" value of availability for these systems in this fleet). The general formulation of this problem is to solve the following for $A_{sys-new}$

$$\sum_{i=0}^{L} \frac{N!}{i!(N-i)!} A_{sys}^{N-i}(1 - A_{sys})^i$$

$$= \sum_{i=0}^{L+1} \frac{(N+1)!}{i!(N+1-i)!} A_{sys-new}^{N+1-i}(1 - A_{sys-new})^i \quad (5.35)$$

For the example above, $A_{sys-new} = 0.922217$. Then the value of a $v\%$ decrease in the availability of a helicopter in the context of the fleet is,

$$C_{\Delta A} = \frac{vC_P}{100(A_{sys} - A_{sys-new})} \quad (5.36)$$

For our example case, if the helicopters cost $C_P = \$20M$ each, then a 1% system availability decrease ($v = 1$), is worth \$7,198,647. Note, the value of availability in this case is application (fleet context) specific (as it should be).

Note that the "marginal" availability analysis in the preceding paragraph is not linear, i.e., the analysis could be done with one fewer ($N = 19$) helicopters and the magnitude of the value from Equation (5.36) becomes \$7,707,129, which is not the same answer.

References

[5.1] Lv, J., Xie, Z., and Xu, Y. (2015). Quantitative relationship between readiness and availability of weapon system and its application, *Proceedings of the International Conference on Modeling, Simulation and Applied Mathematics*, pp. 89–93.

[5.2] Sherbrooke, C. C. (2004). *Optimal Inventory Modeling of Systems*, 2nd Ed. (Kluwer Academic Publishers, USA).

[5.3] Erlang, A. (1948). Solution of some problems in the theory of probabilities of significance in automatic telephone exchanges, in *The Life and Works of A.K. Erlang*, E. Brockmeyer, H. Halstrom, and A. Jensen (eds.) (Transactions of the Danish Academy of Technical Sciences), 2.

[5.4] Cooper, R. B. (1972). *Introduction to Queuing Theory* (MacMillan, USA).

[5.5] Conroy, N., Deane, J. P., and Ó Gallachóir, B. P. (2011). Wind turbine availability: Should it be time or energy based? – A case study in Ireland, *Renewable Energy*, 36(11), pp. 2967–2971.

[5.6] Chief of the Bureau of Naval Weapons (1964). *Handbook Reliability Engineering*, NAVWEPS 00-65-502.

[5.7] Barringer, H. P. (1997). Availability, Reliability, Maintainability, and Capability, Triplex Chapter of the Vibrations Institute. Available at: https://bahan-ajar.esaunggul.ac.id/tkt316/wp-content/uploads/sites/977/2019/11/Pemeliharaan-Rekayasa-Keandalan-Pertemuan-13.pdf.

Problems

5.1 Derive Equation (5.7) starting from Equation (5.4) assuming that $w(t)$ and $g(t)$ are exponentially distributed.

5.2 If the steady-state assumption is not made, derive the availability from Equation (5.4) when $w(t)$ and $g(t)$ are exponentially distributed.

5.3 If the failure rate and the repair rate are exponentially distributed with $\lambda = 6 \times 10^{-5}$ failures per hour and $\mu_{rr} = 5 \times 10^{-2}$ repairs per hour, what is the steady-state (inherent) availability?

5.4 If performing one more preventative maintenance activity per year in the example in Section 5.1.2 results in a reduction in the number of failures per year from 2 to 1.5 (i.e., 3 every 2 years), is there any improvement in the system's operational availability, if so, by how much?

5.5 How do the availabilities in the example in Section 5.1.2 change if there is an additional administrative delay time (ADT) of 20 operational hours that has to be applied to only two of the preventative maintenance activities performed during the 5-year support life of the system?

5.6 Assuming that the times-to-failure and times-to-repair are exponentially distributed, what is the inherent availability of a system consisting of the following three components: Component 1: $\lambda = 0.05$, $\mu_{rr} = 0.067$; Component 2: $\lambda = 0.033$, $\mu_{rr} = 0.053$; and Component 3: $\lambda = 0.04$, $\mu_{rr} = 0.045$? Assume that the components have consistent time units and are connected in series and that all non-failed components continue to operate (accumulate damage) during the time when the failed component is repaired.

5.7 The requirements for a system specify an operational availability of 0.8. The system is intended to operate 24 hours a day, 7 days a week. The system was tested during a 1-month (30 calendar day) period during which the following data was collected: operating time per day $= 8$ hours, 15 total failures were encountered, a total of 60 corrective maintenance hours were spent, 25 total hours was spent waiting for parts and 10 total hours was spent finding the right tools and personnel to perform the maintenance. 5 hours per week as spent doing preventative maintenance. Did the system meet its operational availability requirement?

5.8 Derive the relation in footnote 4 starting with Equation (5.27).

5.9 Consider a digital call processing center. If your center has 5 processing units and each one has an individual availability of 0.89,

(a) If all processing units have to be available for the system to be considered available, what is the availability of this system?

(b) If 3 out of 5 processing units operational is considered sufficient for the system to be operational, what is the availability of the system?

5.10 You are required to create a 36 kW wind farm using 3.6 kW turbines (i.e., the farm is considered to meet its requirements if it
generates 36 kW or more). Assume that the individual availabilities are 0.95 (95%). How many turbines does your farm need
to have? Hint, in this case $(N - L)(3.6) >= 36$ kW, solve for N.

5.11 A defensive missile battery holds 6 missiles. The probability
that an individual missile is non-failed at launch is 90%. What
is the probability that at least 4 missiles are non-failed when a
launch order is issued?

5.12 Show $\Pr(t_m < t_d) = \frac{\mu_{rr}}{\mu_{rr} + \lambda}$ when t_m and t_d are exponentially
distributed.

5.13 For the missile launcher problem (Problem 5.11), what is the
value of a 5% increase in missile availability if each missile costs
$1.5M?

5.14 A system consists of three parts in series (all must work for the
system to work). The three parts have MTBFs of 10, 20 and 30
weeks, respectively. The time necessary to replace the first part
is exponentially distributed with a mean of 1 week, the second
part is 1–3 weeks (uniformly distributed) and the third part has
a gamma distribution with at mean of 10 weeks.

(a) Assuming all parts start good-as-new, what is the expected
time until the first part fails?

(b) What is the probability that part 1 will be the first part to
fail?

(c) What is the expected downtime when the system fails?

(d) What is the inherent availability of this system?

Chapter 6

Sustainment Inventory Management

Inventory is a stock or store of goods or services, kept for use or sale in the future. Inventory analysis exists at two different levels: (1) the inventory of finished goods or end-items that are available for sale or use; and (2) the inventory of parts and materials needed to support future manufacturing and/or the sustainment of systems. Inventory management represents the process of determining how much inventory you have, the state of that inventory, how and where to store the inventory, and under what conditions the inventory needs to be replenished and by how much.

For many kinds of products and systems, inventory is the central aspect of logistics, i.e., the planning, organizing, storing, moving, and accounting for inventory is the basis for nearly all logistics. But, besides being a hedge against supply-chain disruptions, inventory also represents a risk itself because inventory constitutes a significant expense for an uncertain future return. For high-volume commercial products, the inventory management problem is focused on maximizing the financial return on inventory while simultaneously satisfying customer demand and service requirements. For critical systems, which are low-volume, long manufacturing and much longer sustainment life, inventory is also an important attribute of logistics.

The demand for parts (the topic of Chapters 3–5 of this book) is generally the most important input to all types of inventory modeling, but it is only one of several inputs to the inventory problem. In this chapter we will primarily focus on the issues associated with the inventory of parts and materials needed to support the sustainment of critical systems. A brief discussion of end-item inventory management is provided at the end of the Chapter (Section 6.7).

6.1 Inventory Modeling

Inventory modeling determines the optimum amount of inventory that should be held and where the inventory should be located to support production and sustainment processes. This includes addressing questions such as determining the optimum frequency of ordering, deciding on the quantity of parts to be ordered and stored, and tracking the flow of materials so as to avoid delays in delivery.

So, why hold inventory at all? Wouldn't it be optimal (minimum cost) to simply buy each part as it is needed? The answer is yes, and the high-volume product world (e.g., automotive) has embraced this notion through a variety of industry-wide inventory reduction initiatives, including *efficient consumer response* (ECR) and *efficient foodservice response*(EFR) used by the food and grocery industry, *quick response*(QR) used by the textiles industry, *continuous flow manufacturing* (CFM), and *just-in-time* (JIT) in the automotive manufacturing world. However, while these programs successfully manage and minimize many of the costs associated with inventory, e.g., the cost of money, holding costs, shelf life losses, etc., there are significant uncertainties and risks that remain.

Inventory management and inventory management systems are a big business domain for high-volume systems and entire books and courses are dedicated to these topics. The bottom line is that too little inventory (or inventory in the wrong place) can be just as detrimental to an organization as too much inventory.

6.1.1 *Inventory metrics relevant to system sustainment*

There are a myriad of metrics used to measure inventory management. In general, the metrics are an indicator of stock control and they provide insight into how your stock impacts your operations. Inventory metrics also allow inventory operations to be benchmarked and tracked over time.

Many traditional inventory metrics are not particularly useful for the sustainment of critical systems. For system sustainment, the inventory metric(s) that make the most sense boil down to: (1) how much you have (quantity); and (2) how effectively you can respond to inventory demand requests. The key metrics and definitions that appear in inventory measurement and modeling for system sustainment are:

- *Fill Rate*: Fill rate is the fraction of demand that is met through immediate stock availability, without delays or backorders. Fill rate is also referred to as the "issue effectiveness rate". Mathematically, fill rate can be expressed as the probability that a random demand will be filled immediately. In this case there will be a fill (demand satisfied immediately) if the number of items required from stock is k-1 or fewer, so the expected fill rate (EFR) is,

$$EFR(k) = \sum_{x=0}^{k-1} \Pr(x) \qquad (6.1)$$

where k is the number of items in stock (e.g., spares), x is the number of items demanded, and $\Pr(x)$ is the probability that x items will be demanded during a replenishment cycle (possibly the probability of x failures). In the limit as k approaches infinity, $EFR(k)$ approaches 1.

- *Service level*: Service level is the expected probability of not encountering a stock-out during the next replenishment cycle. The cycle duration is the lead time for obtaining a spare. The service level is the probability of being able to service the demand without creating a backorder. Service level is also referred to as the "probability of sufficiency" or the "ready rate".

$$SL(k) = \sum_{x=0}^{k} \Pr(x) \qquad (6.2)$$

Comparing Equations (6.1) and (6.2), $\Pr(k) = SL(k) - EFR(k)$.

- *Backorders*: A backorder is an order for a good or service that cannot be filled at the current time due to a lack of available supply. A backorder may occur because the item is temporarily unavailable, i.e., on order, on allocation (see Section 7.5.3), or in the process of being transferred between inventory locations. Generally, backorders will be fulfilled, but they affect the time required for the demanding system to receive spare parts (i.e., impacting the system's availability). The number of expected backorders was previously given in Equation (5.27) as

$$EBO(k) = \sum_{x=k+1}^{\infty} (x - k)\Pr(x) \qquad (6.3)$$

As an example, consider the following system. Assume that the demand is governed by a Poisson distribution where the mean demand is 5 items ($n\lambda t = 5$). The value of $\Pr(x)$ in Equation (6.1) for different values of x is given by Equation (5.25).[1] Assume that there are 6 spares available ($k = 6$). The fill rate, is given by Equation (6.1) as 0.616, which means that 61.6% of demands will be filled immediately in this case. The service level from Equation (6.2) is 0.762, which means that there is a 76.2% probability of servicing the demand without creating a stockout. The expected number of backorders is found using Equation (6.3) to be 0.493.

Several additional definitions are useful for inventory modeling,

- *Lead Time*: Lead time can take on different meanings depending on the context. In general, it is the time between the placement of an order and receiving the ordered items. Lead time can correspond to placing orders from suppliers to replenish spare parts inventories ("material lead time"), or the time between placing a demand for a spare part with the inventory system and getting the part from the inventory system ("customer lead time").
- *Holding Cost*: Holding cost is the cost of holding inventory. For sustainment this may be articulated as a fraction of the procurement price of the component spent per year to hold it in inventory. For commercial inventory measurement, this is probably articulated as a "carrying cost". In the analysis that follows, holding costs are represented by C_h.
- *Shelf Life*: Shelf life is the length of time an item is allowed to remain in inventory (i.e., in storage) before it cannot be used to manufacture or maintain a system. Shelf life is part, material and

---

[1] This is per replenishment period, meaning that a mean of five demands are received during the period between replenishments of the inventory.

technology specific and depends on the storage conditions and the policies associated with the specific systems for which the inventory is held.

- *Order Cost*: Ordering costs are the costs incurred to create and process an order to a supplier.
- *Cycle Stock*: Cycle stock is the inventory that is intended to meet the expected demand, also called working inventory. See Section 6.2.1.
- *Safety Stock*: Safety stock is the extra inventory kept as a buffer against demand uncertainties. See Section 6.2.2.

6.2 Traditional Inventory Modeling versus Critical Systems Inventory Modeling

For critical systems, we are concerned with ensuring that the required system availability (or readiness) levels are met for a system-of-systems. The question to be addressed is usually some version of how can the logistics be structured or changed to meet a required availability level for less money? Where logistics includes stocking of parts, locating part stocks, locating necessary maintenance resources, etc.

A manufacturing-centric organization is worried about a different logistics problem, namely what is the least expensive way to structure the logistics to avoid disruptions to the manufacturing process. Alternatively, a commercial retailer (e.g., Amazon) might view the problem from a customer satisfaction point of view attempting to maximize metrics like fill rate – the fraction of demands that are met from stock on the shelf. In this case, the penalty for a lack of inventory is a backorder that could result in the loss of a sale (and possibly future sales) from a customer. For critical systems, all unfilled demand is backordered and there are no lost sales, but there could be an availability penalty.

In traditional inventory modeling, optimization of inventory is often performed at the part or item level, where each part in a system is independently addressed. The optimal inventory management problem for critical systems cannot be decomposed into independent optimization problems for each part in the system. For a critical system, value is measured in terms of the availability of the system (or fleet of systems), e.g., part-level optimization does not include inputs for the required system availability, or contractual constraints on the system that may limit the inventory size. Optimal part-level decisions may also be prohibited by system-level resources (namely budget).

6.2.1 *Traditional part-level inventory modeling*

The fundamental questions in part-level inventory modeling are generally: (1) when should parts be reordered (replenishment); and (2) how many parts should be reordered? Consider on classical inventory model referred to as the Economic Order Quantity (EOQ) model where the total cost of parts in the jth period of time is given by

$$C_{Totalj} = C_P D_j + \frac{C_{Or} D_j}{Q} + \frac{C_h Q}{2} \tag{6.4}$$

where

C_P is the purchase price of the part,
D_j is the number of parts needed in period j (demand),
C_{Or} is the cost per order (setup, processing, delivery, receiving, etc.),
Q is the quantity per order, and
C_h is the holding (or carrying) cost per period per part (cost of storage, insurance, taxes, etc.).

The first term in Equation (6.4) is the purchase price (the cost of purchasing D_j parts); the second term is the ordering cost (the cost of making D_j/Q orders in the time period); and the third term is the holding cost (the cost of holding the parts in the time period). In the third term, $Q/2$ is the average quantity in stock (assuming linear part consumption) – this term does not use $D_j/2$ because the maximum number of parts that are held at any time is Q (not D_j).

Equation (6.4) can be used to solve for the quantity per order (Q) that minimizes the total cost of parts in a period of time. To solve for the optimal order quantity, minimize the total cost:

$$\frac{dC_{Totalj}}{dQ} = -\frac{C_{Or} D_j}{Q^2} + \frac{C_h}{2} = 0 \tag{6.5}$$

Solving for Q we obtain

$$Q = \sqrt{\frac{2C_{Or}D_j}{C_h}} \tag{6.6}$$

The optimum cycle length (time between reorders) is given by

$$T_o = \frac{Q}{D_j} = \sqrt{\frac{2C_{Or}}{C_h D_j}} \tag{6.7}$$

Equation (6.6) is known as the Wilson EOQ Model or Wilson Formula.[2] There are hundreds of variations on the basic EOQ model described above. Some of these include volume discounts, loss of items in inventory (physical loss or shelf life issues), accounting for the ratio of production to consumption to more accurately represent the average inventory level, probabilistic demand, transportation costing, and accounting for the order cycle time. In Section 7.5.1.3 we will introduce a simple model for design refresh planning that is essentially an EOQ model.

Critical-system sustainment is characterized by small demands for expensive parts (with large holding costs) spread over long periods of time. For parts in critical systems, C_h is often a function of the part price C_P (it is not uncommon for the holding cost for electronic parts to be 10–20% of the part purchase price per year). When D_j in Equation (6.6) becomes small and C_h is large, the optimum value of Q given by Equation (6.6) tends towards 1 (when rounded up). This makes intuitive sense, if parts are expensive, cost a lot to hold, and the demand for them is low, buy them as needed (which is a problem for critical system sustainment). The difficulties with Equation (6.6) are that: (1) it implicitly assumes that parts are continuously available to be procured, which they are not over the long support life associated with critical systems (see Section 7.5); (2) Equation (6.6) does not include cost penalties that could be imposed if spare parts are not available and availability is often more important than cost for critical systems;[3] and (3) critical systems are repairable, i.e., they

---

[2]The model was developed by F. W. Harris in 1913 [6.1]; however, R. H. Wilson, a consultant who applied it extensively, is given credit for it.

[3]There is a variant of EOQ called EOQB that includes the cost penalty for backorders. See Problem 6.4.

are not consumables. So, Equation (6.6) is an elegant solution that is useful for high-volume manufacturing applications, but doesn't really contribute much to solving inventory management problems for critical systems.

6.2.2 *Safety stock*

Inherently, demand is uncertain. The quantity per order (Q) determined in Equation (6.6) is referred to as "cycle stock" when the demand D represents the mean of the demand distribution. Safety stock is additional buffer stock held to guard against demand and lead time uncertainties. The safety stock quantity is given by [6.2]

$$Q_s = z\sqrt{\frac{\mu_{lt}}{T}\sigma_d^2 + \mu_d^2\sigma_{lt}^2} \tag{6.8}$$

where σ_{lt} and σ_d are the standard deviations of lead time and demand, μ_{lt} and μ_d are means of the lead time and demand, T is the demand time period used and z is the z-score corresponding to the required service level. Equation (6.8) is simply the z-score multiplied by the sum of the squares of the lead time and demand variabilities.

For example, assume we want to meet demand 95% of the time (i.e., we can tolerate stockouts for this item no more than 5% of the order cycles), then $z = 1.64$. Assume that the mean demand/week is 35 parts with a standard deviation of 10 parts. Assume that the mean lead time to get more parts if the inventory runs out is 5 days (with at standard deviation of 12 hours). What safety stock should be used? In this case, Equation (6.8) becomes,

$$Q_s = (1.64)\sqrt{\frac{5}{7}10^2 + 35^2\left(\frac{0.5}{7}\right)^2} = 14.5$$

This result indicates that 15 or more parts should be held as a safety stock.

6.2.3 *Repairable systems inventory*

Repairable inventory is not the same as spares inventory, although both may serve the same purpose, which is to improve the availability

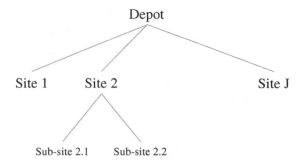

Fig. 6.1. Multi-echelon inventory system.

of equipment in the presence of failures. Spare parts are not necessarily repairable. Repairable inventory theory is more complex than consumable inventory theory – non-repairable spares are consumables.

For the purposes of repairable inventory modeling, the maintenance levels that were introduced in Section 4.1 are generalized into "echelons", Figure 6.1. An echelon represents a level at which inventory is held. In Figure 6.1, the system that the inventory is held for would be located at the lowest level in the tree. For example, if the system is on an aircraft carrier, Sub-site 2.1 could be the part inventory held on the aircraft carrier; Site 2 could be a supply ship; and the Depot would be a port facility. There might even be another level in this case that could be a central inventory facility that supplies several depots. For critical systems it often does not make sense to hold 100% of the inventory of a part at a single location or level. Rather it makes sense to move some inventory closer to the point of need and leave some (maybe the majority) of the inventory in a location that is further from the point of need. Why split up the inventory over multiple echelons? There are lots of reasons: (1) the closer to the point of need you go, the less inventory capacity there is; (2) the closer to the point of need you go, the more expensive the holding costs; (3) inventory that is further from the point of need is less expensive to hold but also takes longer to reach the point of need, thus having a negative impact on availability; and (4) for defense applications inventory that is further from the point of need may be more secure. Echelons may map one-for-one against the maintenance levels articulated in Section 4.1 if inventory is held at those levels.

Sherbrooke [6.3] divides the critical system's inventory problem into the following categories: (1) the multi-echelon problem – tradeoffs between stock at operating locations and their supporting depots; and (2) the multi-indenture problem – tradeoffs between stock for an item and for its subitems. Before addressing multi-echelon (Section 6.4), we will start by discussing single-site inventory models.

6.3   Single-Echelon Inventory Models

A single-echelon inventory model assumes that all the inventory is held at a single distribution point that acts as a central repository between the supplier of the inventory and the field demand for the part. The single distribution point either replenishes its inventory at some fixed interval or when the inventory drops below some threshold.

Typically, the single-echelon problem focuses on optimal replacement quantities for repairable parts that have failed, or have been otherwise lost from the system, or on optimal batch sizes for repair lots. There are two basic approaches to the single-echelon problem: (1) deterministic, batch-procurement (EOQ-type) models (Section 6.2.1); or (2) stochastic, one-for-one replenishment models or batch models. A review of these models appears in [6.4].

Deterministic single-echelon models assume a deterministic demand and known constant recovery rates. EOQ has been extended for repairable systems. Based on Schrady [6.5], the following expression for repairable item total cost can be formulated,

$$C_{Totalj} = C_P D_j (1 - \rho) + C_R D_j \rho + \frac{C_{Or} D_j (1 - \rho)}{Q}$$

$$+ \frac{C_r D_j \rho}{Q_r} + \frac{C_h \rho}{2} \left[Q_r + \left(\frac{1 - \rho}{\rho} \right) Q \right] + \frac{C_{h2} \rho}{2} [Q_r + Q]$$

$$(6.9)$$

where

    C_R is the repair cost per part,
    ρ is the recovery rate (the fraction of failed parts that can be repaired),
    C_r is the fixed cost per repair batch,

Q_r is the quantity per order repair batch, and
C_{h2} is the holding cost per period per repair part.

In Equation (6.9) "repair batch" is the number of items sent for repair at one time, i.e., the items are not sent for repair one-at-a-time, but rather collected into a batch of size Q_r and sent for repair together.

Taking partial derivatives of Equation (6.9) with respect to C_{Or} and C_r, gives,

$$Q = \sqrt{\frac{2C_{Or}D_j(1-\rho)}{C_h(1-\rho)+C_{h2}\rho}} \tag{6.10a}$$

$$Q_r = \sqrt{\frac{2C_rD_j}{C_h+C_{h2}}} \tag{6.10b}$$

Obviously, if $\rho = 0$ (i.e., nothing is repairable), Equation (6.9) reduces to Equation (6.4), and Equation (6.10a) reduces to Equation (6.6). This extension of the deterministic EOQ model to reparable item management is an attempt to balance the number of replaceable and repairable units of an item within an inventory system. Other extensions of EOQ models exist, however, they still represent deterministic (e.g., constant demand), single-item, single-echelon, single-indenture optimizations.

Stochastic, single-echelon models can deal with non-constant demand, i.e., demand that varies over time. In addition, the repair and order lead times (which do not appear in the simple EOQ models discussed so far) are stochastic. Stochastic inventory models require the introduction of the concept of an "inventory policy". The inventory policy is a "rule" that provides a solution to the inventory management process, e.g., whenever the number of parts in inventory drops below x. then buy Q units, where x and Q are the parameters that the policy depends on. So, the goal of stochastic models is to determine the optimal policy and the optimal parameters for that policy.

Stochastic modeling most often means assuming that demand follows a Poisson process and a distribution is used to approximate it. Then equations like Equations (6.4) and (6.9) (and their variations) are solved.

6.4 Multi-Echelon Inventory Models

A sequential single-echelon approach forecasts demand and determines the required inventory for each echelon separately. Multi-echelon inventory optimization determines the correct levels of inventory across the network based on demand variability at the various local sites and the performance (lead time, delays, service level) at the higher echelons. Multi-echelon inventory optimization right-sizes safety stock buffers across the entire supply chain, taking into account the complex interdependencies between echelons, as well as lead times, demand uncertainty, and supply volatility. Simple multi-echelon models generally assume that the echelon structure (Figure 6.1) is a tree-like structure.

The simplest multi-echelon problem is composed of a two-level system with a number of locations (sites) being served by a single central depot. Failures create a demand for items, which may be spared from stock at the local sites; concurrently, the failed item will be placed into the repair at the local site (if possible) or at the depot-level repair facility (if it can't be repaired at the local site). In the event that insufficient stock is available at the local-level, a backorder occurs and demand must be satisfied from the inventory at the depot, either from existing stock or from items completing the repair process.

The problem articulated in the last paragraph, involves setting repairable inventory levels at each of j sites such that a prescribed level of service is achieved (possibly a desired protection level, see Section 4.4.1.1), given repair rates, transportation times and budget constraints. If condemnations (permanent removal of items that are deemed unrepairable) are allowed, then the model must plan for procurement quantities and timing. Additional complications may arise such as allowing transfers of units from one site to another to satisfy demand (transshipment), cannibalization, re-distribution of repaired units among the sites, capacity constrained repair facilities, batching of repair and replenishment quantities, and other factors.

6.4.1 *METRIC*

Many of today's repairable inventory models are based on the METRIC (Multi-Echelon Technique for Recoverable Item Control) model, which was developed by Craig Sherbrooke in the late 1960s at the

> ## Washington Metro's Multi-Echeloned Support Network [6.6]
>
> Consider the Washington Metro – the rapid transit system servicing the Washington DC metropolitan area. As of November 2016, Metro was operating a fleet of over 1200 railcars composed of six major design series from five different major manufacturers. Although the railcars are externally similar in appearance, the different railcar series share varying degrees of component commonality. Metro stores and maintains their railcar fleet at nine rail yards located throughout the system, and component overhauls are performed either by Metro-internal overhaul shops or by outside vendors. Metro's Office of Procurement and Materials "manages an inventory of tens of thousands of items at the [Rail] Car Maintenance operating locations as well as the [central] Metro Supply Facility" [6.6].

RAND Corporation for the U.S. Air Force [6.7]. The METRIC model is designed to set repairable inventory levels and allocate the inventory so as to achieve some desired level of service, measured by expected backorders at the local level.

METRIC takes a systems view of the repairable inventory problem since it is concerned with setting initial inventory levels and the distribution of the inventory amongst multiple local locations in order to minimize backorders system wide. There are several key simplifying assumptions in METRIC:

- The site (local) level is resupplied from the depot, not from another local site (i.e., no lateral resupply).[4]
- There is no condemnation of parts (i.e., all parts are repairable). Also, no new systems that could demand parts are added (i.e., the population of systems is static).
- All parts are the same.
- Q (the EOQ) is 1, Q_r is also 1 (i.e., there is no batching of repair units).
- Steady-state (average) inventory levels are determined.

---

[4]Lateral resupply would make the echelon structure non-tree-like.

The average annual demand for parts on a depot that supports J local sites is given by

$$m_0 = \sum_{j=1}^{J} m_j(1 - f_{rj})$$

where

m_j is the average annual demand for parts at site j (all of the demands are for repairs), and

f_{rj} is the fraction of repairs that can be made at site j (repairs that cannot be made at the site have to be made at the depot).

The average number of parts under repair at (or being resupplied to) site j is given by $(j > 0)$,[5]

$$\mu_j = m_j \left\{ f_{rj}T_j + (1 - f_{rj})\left(O_j + \frac{EBO[s_0|m_0T_0]}{m_0}\right)\right\} \qquad (6.11)$$

where

T_j is the average time per repair for a part at site j,

O_j is the order time (ordering and shipping to the site from the depot) for site j,

$EBO[s_0|m_0T_0]$ is the expected number of backordered spare parts (demand requests that cannot be filled from existing stock) at the depot (the "0" subscripts indicate the depot as opposed to site j),

m_0T_0 is the average depot pipeline, and

s_0 is the depot stock (inventory) level.

This model does not actually distinguish between repair and in-pipeline times, i.e., T_j and O_j represent a combination of repair, order and other in-pipeline times. Equation (6.11) reduces to m_jT_j if $f_{rj} = 1$ (all parts repaired at site level), if for example, $m_j = 23.2$

---

[5]Equation (6.11) is commonly called the average pipeline because it denotes the number of parts in the pipeline being repaired at the site or resupplied to the site from a higher echelon. Generally, pipeline refers to the number of parts that have been ordered but not received. The concept is that failed parts enter the repair cycle (the pipeline) and the variability of pipeline's size and turnaround time are the focus of inventory optimization. METRIC and similar models are pipeline optimization algorithms.

demands/year, $T_j = 0.01$ years/site repairable part, and $f_{rj} = 1$ site repairable parts/demand at the site then $\mu_j = 0.232$ or there are an average of 0.232 repairs occurring at any given time at site j (note μ_j is a dimensionless number). If $f_{rj} = 0$ (all parts shipped to the depot for repair) and there are no backorders, Equation (6.11) reduces to $m_j O_j$. If for example, $m_j = 23.2$ demands/year, $O_j = 0.02$ years/site repairable part, and $f_{rj} = 0$ then $\mu_j = 0.464$ or there are an average of 0.464 parts being resupplied to site j at any given time.

The expected number of backorders is given by Equation (6.3). Order cost and holding cost are not included in this model because all cases are one-for-one replenishments, which defines the number of orders and the average stock level.

What is going on in Equation (6.11). This model assumes that there is no condemnation of parts, i.e., no parts are retired from the system and every part can always be repaired either at the site or the depot. As repairable parts fail (creating demands for repair), they are replaced with working spare parts from inventory. The failed parts are then repaired either at the site or at the depot and after a time equal to μ_j/m_j they are returned to the inventory (stock) either at the site or depot. When a repair demand arrives and there is no spare available at the site or depot, a backorder happens. If there are no spares stocked at the depot or the site, then μ_j is the expected number of backorders.[6]

Let's work an example (using example values from [6.3]), assume the following supply system:

$m_j = 23.2$ demands/year at each site j,
$T_j = 0.01$ years/site repairable part to repair at site j,
$f_{rj} = 0.2$ (20% of the site j demands that can be repaired at site j),
$O_j = 0.1$ years/part for site j to order from the depot,
$T_0 = 0.02531$ years/part to repair at the depot, and
$J = 5$ identical local sites.

---

[6]This brings up the question of the difference between the pipeline (μ) and the expected number of backorders (EBO). A backorder is a demand that can't be fulfilled. The pipeline could be larger, equal to, or smaller than the number of backorders. If there are no spares and no condemnation of parts (all parts are always repairable), then the number in the pipeline exactly equals the number of backorders.

What do we wish to determine? First, we will determine the expected number of backorders for any level of stock at the depot and/or the sites. Armed with the expected backorder levels, we will determine the optimum allocation of a given stock level across the sites and the depot, i.e., how many to put where.

From Equation (6.11) the average annual demand at the depot for repair is $m_0 = (5)(23.2)(1-0.2) = 92.8$ parts/year. The expected number of backorders at the depot if the depot stock is $s_0 = 0$ is given by solving Equation (6.3) with $\lambda t = m_0 T_0 = 2.349$ and $k = 0$, which gives $EBO(s_0 = 0) = 2.349$ parts (this is the expected backorder level at the depot if the stock level at the depot is 0). Now we can compute the average number of parts under repair at each site using Equation (6.11), which comes out to be $\mu_j = 0.7017$. This means that there are an average of 0.7017 parts under repair at each site if the depot stock level and the site stock levels are all 0. Multiplying μ_j by the number of sites (J) gives, 3.5087, which is the total number of backorders for the system consisting of the depot and 5 sites if there is no inventory at the depot or any of the sites.

The depot backorders for other values of s_0 (assuming no stock at the sites) can be computed using the same process with $k = s_0$ and finding the corresponding μ_j values from Equation (6.11), then multiplying those values by J. For example, $EBO(s_0 = 1) = 1.4443$, $\mu_j = 0.5209$, and the expected backorders for 1 part in stock at the depot and no inventory at the sites is 2.6043. This process produces the values in the first column of Table 6.1.

What if there are parts stocked at the sites? The number of backorders at the sites is can also be computed from Equation (6.3) with $\lambda t = m_j T$ where T is the average time for a site repair request to be resolved corresponding to the depot stock level,

$$T = \frac{\mu_j}{m_j} = \left\{ f_{rj} T_j + (1 - f_{rj}) \left(O_j + \frac{EBO[s_0 | m_0 T_0]}{m_0} \right) \right\} \quad (6.12)$$

For $s_0 = 0$, $T = 0.0302$ years, giving $\lambda t = m_j T = 0.7017$ demands. The expected backorders for each site from Equation (6.3) with $k = 1$ are $EBO(s_j = 1) = 0.1975$ parts. The total expected number of backorders if n of the J sites have 1 part in inventory (and the depot has no parts) is given by

$$EBO = EBO(s_j = 1)n + (J - n)EBO(s_j = 0) \quad (6.13)$$

Table 6.1. Expected backorders for up to 8 parts in stock anywhere in the system.

Depot Stock	Total Site Stock								
	0	1	2	3	4	5	6	7	8
0	3.508768	3.004483	2.500199	1.995914	1.491629	0.987344274	0.830929	0.674515	0.5181
1	2.604255	2.19827	1.792284	1.386299	0.980314	0.574329021	0.477737	0.381145	
2	1.924018	1.604602	1.285186	0.965771	0.646355	0.326939332	0.269415		
3	1.507167	1.246924	0.986681	0.726438	0.466195	0.205952434			
4	1.296527	1.068115	0.839702	0.611289	0.382876				
5	1.206973	0.992505	0.778036	0.563568					
6	1.1743	0.964981	0.755663						
7	1.163892	0.956221							
8	1.160956								

Equation (6.13) generates the first row of Table 6.1. If the expected backorders are generated for all the combinations of up to 8 parts in inventory for this system, one obtains Table 6.1.

Each diagonal in Table 6.1 represents a constant total number of parts in the inventory system (distributed all possible ways). The minimum backorder in that diagonal (shaded in Table 6.1) represents the optimum allocation of those parts between the depot and the sites. For example, for 5 parts in inventory, the optimum is 3 parts at the sites and 2 at the depot (the expected backorder for this case is 0.965771).

Finally, the multi-echelon problem is addressed using marginal analysis, i.e., the next investment is allocated to the part that produces the largest decrease in the expected backorders divided by its cost.

Many extensions of the METRIC model have been published. MOD-METRIC, for example, allows multiple levels of indenture [6.8]; VARI-METRIC addresses multi-indentured systems and variance of pipeline inventories [6.9][7]; and Dyna-METRIC accommodates surges in demand [6.10]. There are many other variations of the basic MET-RIC model, see [6.4] for a good review.

6.5   Cannibalization and Salvage

Cannibalization is the practice of taking usable parts from one item to complete the repair of another item. Salvage is similar to cannibalization and means that a discarded item (presumably discarded because it has failed and cannot or will not be repaired), can be salvaged for specific parts that can be used to compensate for failures. In some cases, a particular instance of an asset is set aside and becomes a source of cannibalized spares for other assets, i.e., a "hanger queen" or "cann-bird". In other, more egregious cases, the last system in the door for repair or refurbishment is cannibalized to complete the repair of the systems that has been in the facility the longest.

---

[7]The METRIC approach can result in an underestimation of the backorders. VARI-METRIC takes into account not only the mean pipeline, but the also the pipeline variance.

While cannibalization is supposed to be a "last resort" inventory of parts, it is often relied upon and is a not-uncommon occurrence near the end of system life, i.e., when you need just a few more parts to make it to the end-of-support.

Sometimes cannibalization is viewed as an indicator of the failure of an inventory management system. While it is true that the failure of the inventory system will lead to more cannibalization, cannibalization by itself does not necessarily mean that the system has failed. In some cases, cannibalization represents the most efficient path to increase readiness and availability, and decrease costs.

A good review of cannibalization appears in [6.11].

6.6 Shelf Life

Shelf life is defined as the length of time an item may be stored without becoming unsuitable for use.[8] The shelf life of items in inventory is based on a number of factors, including the type of item, the type of materials used in the item, the application that item is used within, and the environmental conditions in which the inventory is held. In some cases, inventory for critical systems must be held for a very long time (decades) in inventory before it is used.

A significant body of work exists modeling inventory systems that include shelf life. Most applications are in the perishable goods space (i.e., food and health care). Treatments include modifications to the EOQ model (Section 6.2.1), Markov chain models, and a host of other types of models. The EOQ model developed in Section 6.2.1 can be modified to include shelf life. If the shelf life of the part being ordered is T_s (measured in the same time periods as the demand, D), then $Q < D_j T_s$. More generally, shelf life is modeled as a continuous deterioration of items in the inventory using an inventory depletion model like,

$$\frac{dI(t)}{dt} + \theta(t)I(t) = D(t) \tag{6.14}$$

---

[8]Shelf life is not the same as loss in inventory. Loss in inventory refers to the portion of the inventory that is either physically lost (misplaced, inadvertently disposed of, misclassified, etc.) or damaged/contaminated so as to be unusable. Alternatively, the term "condemnation" is used to describe subsystems that are determined to be unserviceable.

where

$I(t)$ is the inventory quantity as a function of time,
$\theta(t)$ is the deterioration rate as a function of time, and
$D(t)$ is the demand rate as a function of time.

The second term in Equation (6.14) is the safety stock. Equation (6.14) can be solved under different assumptions. A shelf life that is determined by policy creates a deterioration rate that is a step function, i.e.,

$$\theta(t) = 1 - u(t - T) \qquad (6.15)$$

Where $u()$ is a unit step function. In this case $\theta(t)$ is zero when $t < T$ and 1 thereafter.

6.7   End-item Inventory

End-items are defined as a finished good or saleable product. In the context of this book, an end-item is a complete system that is not used as spares for higher-level systems, e.g., aircraft, trucks, buses, etc. While the concepts developed in this chapter are applicable to all types of items, including end-items, it is useful to look exclusively at end-items as well.

In the context of end-items, inventory means counting readiness reportable items. One way to breakdown this inventory count is the following [6.12]:

$$TOAI = TAI + TII + EPSLI \qquad (6.16)$$

where

$TOAI$ is the total overall asset inventory,
TAI is the total active inventory,
TII is the total inactive inventory, and
$EPSLI$ is the end of planned service life inventory.

The TAI is a function of the primary asset inventory (mission capable), the backup inventory, the prepositioned inventory, and the attrition reserve. The TII consists of the bailment-contract. Lease, loan, training assets, reconstitution-preserved assets, and other storage assets. The $EPSLI$ are the assets on the property book that are never coming back into service. The *TAI* are the assets that contribution to the operational availability (A_o).

References

[6.1] Harris, F. W. (1913). How many parts to make at once, *Factory, The Magazine of Management*, 10(2), pp. 135–136, 152.

[6.2] Simchi-Levi, D. and Kaminsky, P. (2003). *Designing and Managing the Supply Chain: Concepts, Strategies, and Case Studies*, 2nd Ed. (McGraw-Hill, USA).

[6.3] Sherbrooke, C. C. (1992). *Optimal Inventory Modeling of Systems, Multi-Echelon Techniques* (Wiley, USA).

[6.4] Guide Jr., V. D. and Srivastava, R. (1997). Repairable inventory theory: Models and applications, *European Journal of Operational Research*, 102, pp. 1–20.

[6.5] Schrady, D. A. (1967). A deterministic inventory model for reparable items, *Naval Research Logistics Quarterly*, 14(3), pp. 391–398.

[6.6] Transportation Resource Associates (2007). *WMATA Triennial On-Site Safety Review Final Report*, Available at: http://www.tristateoversight.org/pdf/program/TOC%202007%20Triennial%20Review%20Final%20Report.pdf.

[6.7] Sherbrooke, C. (1968). METRIC: A multi-echelon technique for recoverable item control, *Operations Research*, 16, pp. 122–141.

[6.8] Muckstadt, J. A. (1978). Some approximations in multi-item, multi-echelon inventory systems for recoverable items, *Naval Research Logistics Quarterly*, 25, pp. 377–394.

[6.9] Sherbrooke, C. (1986). VARI-METRIC: Improved approximations for multi-indenture, multi-echelon availability models, *Operations Research*, 34, pp. 311–319.

[6.10] Hillestad, R. J. (1982). Dyna-METRIC: *Dynamic Multi-Echelon Technique for Recoverable Item Control*, R-2785-AF, Rand Corporation.

[6.11] Tsuji, L. C. (1999). Tradeoffs in Air Force Maintenance: Squadron Size, Inventory Policy, and Cannibalization, M.S. Thesis, Technology and Policy, MIT.

[6.12] Herzberg, E. (2019). The sustainment enterprise metrics dashboard – status, value-added, and way forward, *Proc. DoD Maintenance Symposium*. Available at: https://www.sae.org/binaries/content/assets/cm/content/attend/2019/dod/2019-presentations/bo9—sustainment-dashboard—herzberg.pdf.

Problems

6.1 Show that in the limit as k approaches indefinitely, the $EFR(k)$ given in Equation (6.1) approaches 1.

6.2 For the numerical example in Section 6.1, if there was a fleet of 90 aircraft and the item was the only item requiring spares, what would the supply availability be? Hint, supply availability is defined in Section 5.2.1.

6.3 A production system consumes 1200 of a ceramic capacitors per month. It costs \$16 to place an order for the capacitors with its supplier. The price of each capacitor is \$0.75. The holding costs for the capacitors are 20% of the price of the capacitor per month. What is the optimal order quantity? How often should the capacitors be ordered?

6.4 There is a variant of the EOQ model called EOQ with Backorders (EOQB). This variant allows for stockout situations that have associated backorder costs. What is the applicability of the EOQB model to critical systems where a backorder cost is much larger than the part holding cost?

6.5 In Problem 6.3, assume that the standard deviation on the monthly capacitor demand is 121 capacitors and the lead time to obtain more capacitors if you run out is 3 days (with a standard deviation of 6 hours). If you want to avoid stockout situations 90% of the time, how much safety stock should you have? Assume 30 days per month.

6.6 Derive Equation (6.9) and show that Equations (6.10a) and (6.10b) are the quantities associated with the optima.

6.7 For the example in Section 6.4.1, compute $EBO(s_0)$ for s_0 values of 2 through 8 using Equation (6.3).

6.8 Find the expected backorders for the example in Section 6.4.1 when there are 9 total spares in the inventory system, i.e., generate the 9th column and 9th row of Table 6.1.

Chapter 7

Supply-Chain Management

Supply chains are commonly defined as the network between a company and its suppliers used to produce, distribute and sustain a specific product to a final customer.[1] The supply chain can also represent the sequence of steps necessary to combine or transform a set of components into a system for the customer. This network includes many different activities, people, organizations, information, and resources.

Supply chains are often represented graphically as a network that describes the relationships between the elements of the supply chain, Figure 7.1. The vertical slices of the supply chain are commonly referred to as "tiers" where tier 1 might represent all the suppliers that the system manufacturer sources directly from, tier 2 are the suppliers that the tier 1 supplier's source from, etc. The links between suppliers represents the flow of goods and services (as well as the flow of money, intellectual property and data).

The term "supply chain" infers a linear or serial system in which each tier only has one supplier, but, for complex systems the supply chain is really a network.

---

[1]Note, the majority of definitions in the business world confine supply chains to just "produce and distribute", i.e., they often omit "sustain", but for critical systems, supply-chain management is potentially more challenging in the sustainment phase of the system's life cycle.

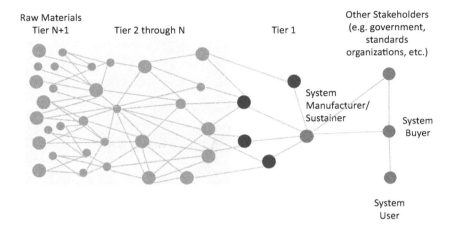

Fig. 7.1.   Multi-tiered supply-chain.

7.1   The Supply Chain for Critical Systems

There are several supply-chain issues associated with sustaining critical systems that are not generally considered (or don't represent a priority) in the broader manufacturing and distribution-centric supply-chain management space:

- Critical-system supply chains must persist for long periods of time (decades).
- Critical systems are safety-, mission- and infrastructure-critical, therefore the failure of the supply chain can have catastrophic outcomes, i.e., supply-chain risk management takes on a different complexion.
- Critical systems often do not "own" (control or drive) significant portions of their supply chains.

Supply-chain disruption and compromise is a significant problem for all types of products, but for critical systems, supply-chain problems are compounded and can create significant dangers that are not commonly experienced by commercial products. First, critical systems are expensive to develop and procure – often requiring investment (possibly of taxpayer money) of hundreds of millions of dollars, e.g., airplanes, subway trains, 911 systems, power plants, military systems, etc. Because they are so expensive to procure, they must last a long time. You may upgrade your cell phone every 2 years, but airlines fly aircraft for 30+ years, and the military is planning to fly some aircraft well beyond 50 years.[2] Secondly, critical systems are often "safety-critical" systems that are highly qualified and certified. This means that the systems have extremely limited flexibility to change, e.g., when a part fails, the system has to be repaired with the same parts it was originally manufactured with; repairing with a newer version of the part could require a lengthy (and very expensive) qualification and certification process that may be impractical.[3] Third, critical systems are not generally a supply-chain driver, which simply means that critical systems are forced to rely on the same supply chain as high-volume commercial systems that are manufactured and supported for much shorter periods of time. For example, your cell phone has a dedicated supply chain the exists primarily for it, but the electronics in an aircraft does not. The majority of the electronic parts in your cell phone have an "in-market" life of 2–3 years, after which the parts are discontinued by their manufacturers in favor of newer versions of the parts. So, how do you support an airplane for 30+ years using parts that were obsolete (i.e., no longer available from their original manufacturers) before the first plane was delivered to its first customer?

Critical systems suffer from supply-chain aging, i.e., supply chains simply grow old or evolve away from critical-system customers. Supply chains follow the highest volume market, often leaving lower-volume markets behind, and if the lower-volume markets cannot adapt to the direction that the supply chain moves, they will encounter obsolescence problems, i.e., an inability to source the items that they need. This form of supply-chain aging creates an insidious set of risks for critical systems.

---

[2]The USAF's B-52 could see service well into the 2040s, at which point, some of the airframes will be approaching 90 years of age. Generally, taxpayers often won't approve the budget to replace infrastructure until the old infrastructure is unusable or a catastrophic event occurs.

[3]Today, 20% of the power generation in the United States is from nuclear. The majority of the electronics technology in nuclear control rooms is analog, not digital. Because there is no practical re-certification path to update these systems, they have to continue to be supported with long-obsolete electronic components.

For critical systems, supply-chain risks are a more pervasive problem during the sustainment phase of the system (as opposed to manufacturing). Most critical systems are referred to as "sustainment-dominated" (see Section 1.3). Therefore, supply-chain risks are a bigger problem during the long sustainment life (decades) than during the manufacturing of the system – this changes the nature and management of the supply-chain risks.

7.2 Supply-Chain Risks

Before focusing on the particular supply-chain risks encountered by critical systems, let's categorize the risks. Supply-chain risks have been classified many different ways to suit a myriad of different needs. In Figure 7.2, we categorize risks based on their source (motivated by [7.1]) – either internal to the organization (endogenous) or external (exogenous), and their intent (intentional or inadvertent).

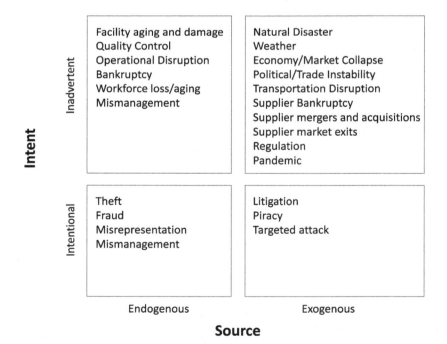

Fig. 7.2.   Taxonomy of supply-chain risks.

It is also important to distinguish supply-chain disruptions from compromise:

- *Disruptions*: Disruptions are defined as any unforeseen event that disturbs the normal flow of goods and materials in a supply chain. Disruptions are caused when the customer fails to receive the necessary parts (or other resources, e.g., software, data, or service) from their supply chain on the required schedule and may result in the inability to manufacture the system, sustain the system, or otherwise obtain a required outcome from the system. Disruptions are dynamic in the sense that they have a probability of occurrence in a period of time and they have a variable duration.
- *Compromises*: Compromise impacts the quality and/or security of a final system. Compromise occurs when the customer receives parts (or subsystems) that are compromised and those compromised parts lead to a compromised system. In this case "part" could refer to hardware, software, data, algorithms, etc.

It is important to distinguish disruptions from compromises because the strategies for managing them can be substantially different. For example, storing excess inventory (buffering) to protect against a catastrophic event can alleviate the impact of the event, but buffering can exacerbate the negative impact of a compromise event [7.1], i.e., the buffered parts may be compromised. However, it is also important to understand that disruptions and compromises are not independent of each other. Compromises can cause disruption, i.e., the detection of a compromise by a member of the supply chain may (depending on the motivations of the agent that detects the compromise) cause a disruption in the flow of parts until the compromise can be localized and resolved. Alternatively, disruptions in the supply chain can cause suppliers in high tiers of the supply chain to search for alternate suppliers thus increasing the risk of procuring a compromised part.

7.2.1 *Propagation of supply-chain risks*

Tactical supply-chain management deals with how a risk is mitigated when it occurs. Strategic supply-chain management focuses on being able to forecast a supply-chain problem and potentially act before it arrives, or alternatively, making an existing supply chain more resilient to disruptions and compromises. The question addressed in forecasting supply-chain problems is how disruptions and compromises in a multi-tiered supply chain propagate. Ultimately, one must be able to predict the following if one or more of the participants in a supply chain are disrupted or compromised,

- Will the final customer be disrupted? i.e., does a disruption "snowball" [7.2]?
- If the final customer will be disrupted, when will that occur (how long after the disruption or compromise)?
- If the final customer will be disrupted or compromised, what will the duration of the disruption or compromise be?
- What is the probability that the final product will be compromised?
- What form will the disruption or compromise take if it reaches the final customer?

The "final customer" could refer to the organization taking delivery of the final product (sometimes called the "operator"), or any other stakeholder in the supply chain who is trying to understand the risks associated with their ability to deliver their outcome undisrupted and uncompromised to their customer. Where "outcome" could be hardware, software, data, services, or a combination of any or all of these.

7.3   Part Sourcing Strategies

Procurement is the process of obtaining the materials you need. Sourcing involves finding and vetting (qualifying) the suppliers of those materials. Conventionally, sourcing strategies are about finding the balance between the part price and part availability that is appropriate for your organization.

Numerous approaches for optimal sourcing for parts exist, however, the existing analyses are generally part procurement-price centric, which is appropriate for high-volume consumer products. Strategic sourcing for these sorts of products often focuses on split-award auctions and similar approaches. In a split-award auction, the buyer announces prior to the procurement auction, that the best bidder will be allocated the highest share of the contract, the second-best bidder will be allocated the second-highest share of the contract, and so on thus creating as much price competition as possible (the lower a supplier bids, the more business volume they get).

Sourcing strategies, like split-award actions, don't work very well for low-volume system – there simply isn't enough business to spit amongst competing suppliers (and price often isn't the most important attribute that suppliers compete on, e.g., they might compete on lead time). In addition, qualifying suppliers (and keeping them qualified over time) is expensive (the fewer the better), but fewer suppliers increases the risk of disruptions in the availability of parts. The sections that follow define and discuss sourcing strategies that are commonly considered for critical systems.

7.3.1 *Sole sourcing*

Sole sourcing means that all parts are procured from a single source because there is only one source. Sole sourcing is rarely employed as a definitive sourcing strategy but rather, it is imposed by the market. Sole sourcing may be necessary early in the part's life cycle or as a part approaches obsolescence. For some "state-of-the-art" technologies, the required part may only be available from one specialized manufacturer. A change in the supply-chain structure may occur in order to use additional suppliers when and if they emerge. Sole sourcing may also be observed following the use of other sourcing strategies as a customer may revert to sole sourcing to utilize the last remaining supplier in the market.

7.3.2 *Single sourcing*

Single sourcing is an exclusive relationship between an original equipment manufacturer (OEM) and a single supplier. In the case of single sourcing, only one source is used by choice (not necessity).

A single-sourcing strategy is only possibly when more than one supplier exists in the market, a situation that is common as a part begins to mature and prior to a part phasing-out of the market place. The part and supplier selection is usually a result of a "winner-take-all" (WTA) auction. A WTA auction is a procurement process that involves selecting and awarding a contract to a winning supplier based on purchase price bids for a specific part. Single sourcing promotes strong customer-buyer relationships by providing opportunities for both parties (supplier and customer) to streamline procurement practices, reduce lead times, and reduce inventory. The customer-supplier relationship also allows part and product decisions to be coordinated. An added benefit of single sourcing is reduced prices for large quantity purchases as a result of "economy of scale". Unfortunately, while single sourcing facilitates better customer-supplier relationships and can reduce multi-supplier qualification costs, the susceptibility to supplier-specific disruptions is greater, i.e., a single point of failure is created. Although single sourcing puts the buyer at risk due to the probability of supply-chain disruptions or failures, there may be sufficient confidence in the supplier's ability to warrant its use. Single sourcing is often used when tooling costs (or source qualification costs) are high and only one supplier is willing to commit to manufacturing a particular part.

7.3.3 *Dual sourcing*

Selection of a primary source and a secondary source using a split-award auction or similar mechanism. In this case, parts are sourced from both. The primary contract accounts for a majority of the required parts and is intended to maximize "economy of scale" while the secondary contract is expected to be more flexible to account for demand uncertainty. From a strategic perspective, a secondary supplier offers supply redundancy should the primary supplier fail to meet the OEM's demand for parts.

7.3.4 *Second sourcing*

Second sourcing involves purchasing parts from a primary supplier while maintaining a backup/secondary supplier. This sourcing strategy potentially decreases the impact of disruptions since part procurement needs can be rerouted to the second supplier when the primary supplier is disrupted (not able to supply parts). However, while second sourcing is good for supplier negotiations (manufacturers can put pressure on the price), additional qualification and support costs can negate its benefits.

7.3.5 *Multi sourcing*

Multi-sourcing is the use of multiple suppliers simultaneously or as a combination of one or more sourcing strategies. The use of multiple suppliers may involve a combination of attributes from all other sourcing strategies. Companies implement various multi-sourcing strategies to prepare for imminent and unpredictable events. To counteract the likelihood of supply-chain disruptions, many organizations prefer multi-sourcing as a way to distribute part purchases among more than two suppliers should one or more suppliers fail to meet required purchase volumes.[4]

7.3.6 *Parallel sourcing*

In parallel sourcing, identical (or very similar) parts are procured from multiple suppliers where parts from each supplier are segregated and used in a unique products rather than being picked for use from a pool of common parts. Therefore, parallel sourcing offers the benefits of multi-sourcing (lower procurement cost from competitive pricing and contract auctions) while allowing commonality across product designs. Parallel sourcing resembles single sourcing in all aspects except that the parts obtained from a particular supplier is limited to a single product. This strategy limits the consequences associated with a specific supplier-related supply-chain disruption but may still be susceptible to common-mode supplier failures or disasters.

7.4 Part Buffering

Buffering is defined as the storage of an inventory of parts that is equal to the forecasted part demand (for both manufacturing and maintenance requirements) corresponding to a fixed future

---

[4] A risk that can appear as supply chain's age is that what appears to be multiple sources for a part, is really just a single source. If the supply network is illuminated through several tiers, one might find that the sources of risk for one supplier are identical to another supplier in which case sourcing from multiple suppliers does little to mitigate the risk of disruption.

time period. Buffering is a common proactive mitigation strategy employed to offset the impact of disruptions. Buffering (also referred to as "hoarding") is a type of dynamic inventory policy. Buffering violates the basic tenants of lean manufacturing, which seeks to minimize the need for holding and managing large inventories of parts. Lean manufacturing approaches, however, assume that suppliers can provide parts without interruption over long periods of time. Long duration disruptions in supply can be extremely problematic for low-volume long-life systems that depend on parts when lean manufacturing approaches are used.

When a supply disruption occurs, new parts are no longer being delivered and the production and support begins to rely on the buffered inventory. However, if the disruption extends past the buffering duration, parts are backordered. The backorder quantity is the number of parts on backorder at the end of the disruption period (see Section 5.2.1).

While buffering can be shown to significantly decrease the penalty costs associated with disruption events, there are also negative impacts that need to be considered. Obviously, buffering has a larger capital cost than buying parts as they are needed (see Problem 7.1). Also, buffering (if left unchecked) can delay the discovery of counterfeit parts in the inventory (i.e., propagating a compromise). Similarly, long-term storage of parts can lead to part deterioration (e.g., the reduction of important solderability characteristics for electronic parts if they are not stored correctly). For these reasons, OEMs that utilize long-term buffering as a disruption mitigation strategy may need to regularly assess the status/condition of buffered parts, and employ first-in-first-out type policies.

Lifetime buy, which is addressed in Section 7.5.1.2 is a form of buffering that occurs when a part's procurement window closes, i.e., when the part will never be procurable again in the future.

7.5   Aging Supply-Chain Challenges

Critical systems are subject to all of the same supply-chain risks as high-volume products. But, in addition, critical systems, because of their long life cycles and low volumes, suffer from a myriad of "aging supply chain" issues, which are discussed in the following subsections.

7.5.1 *Obsolescence*

Technology obsolescence is defined as the loss or impending loss of original manufacturers of items or suppliers of items or raw materials [7.3]. The type of obsolescence addressed in this section is referred to as diminishing manufacturing sources and material shortages (DMSMS), which is the result of the unavailability of technologies or parts that are needed to manufacture or sustain a product. DMSMS means that due to the length of the system's manufacturing and support life and possible unforeseen life extensions to the support of the system, the necessary parts and other resources become unavailable (or at least unavailable from their original manufacturer) before the system's demand for them is exhausted. Part unavailability from the original manufacturer means an end of support for that particular part and an end of production of new instances of that part (i.e., the part is obsolete).[5]

The DMSMS-type obsolescence problem is especially prevalent in "sustainment-dominated" systems that have long enough design cycles that a significant portion of the technology in them may be obsolete prior to the system being fielded for the first time, as shown in Figure 7.3. Once in the field, the operational support for these systems can last for twenty, thirty or more additional years. A possibly more significant issue is that the end-of-support date for systems like the one shown in Figure 7.3 is not known and will likely be extended beyond the original plan one or more times before the system is retired.

For systems like the one shown in Figure 7.3, simply replacing obsolete parts with newer parts is often not a viable solution because

---

[5]There are several other types of "obsolescence" that appear in the operations research and business literature, which are not DMSMS-type procurement obsolescence. Inventory or sudden obsolescence refers to the opposite problem from DMSMS obsolescence [7.4]. Inventory obsolescence occurs when the product design or system specifications changes such that existing inventories of components are no longer required. Alternatively, planned obsolescence is a policy followed by the manufacturer of a consumer product to obsolete the current products with newer versions (and stop supporting old versions) so as to force the customer to continuously upgrade their item. Pseudo-functional obsolescence is planned obsolescence which appears to introduce innovative changes into a product, but in reality, does not.

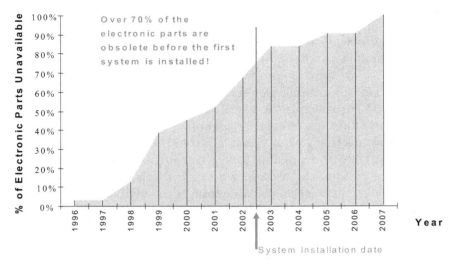

Fig. 7.3.   Percent of COTS parts that are un-procurable versus the first 10 years of a surface ship sonar system's life cycle (courtesy of NAVSURFWARCENDIV Crane).

of high re-engineering costs and the potentially prohibitive cost of system re-qualification and re-certification. For example, if an electronic part in the 25-year old control system of a nuclear power plant fails, an instance of the original component may have to be used to replace it because replacement with a part with the same form, fit, function and interface that isn't an instance of the original part could jeopardize the "grandfathered" certification of the plant.

Sustainment-dominated products particularly suffer the consequences of electronic part obsolescence because they have no control over their part supply chain due to their relatively low production volumes. DMSMS-type obsolescence occurs when long field life systems must depend on a supply chain that is organized to support high-volume products. Obsolescence becomes a problem when it is forced upon an organization; in response, that organization may have to involuntarily make a change to the product that it manufactures, supports or uses.[6]

---

[6]Researchers who study product development characterize different industries using the term "clockspeed", which is a measure of the dynamic nature of an industry [7.5]. The type of industries that generally suffer from DMSMS problems

The long-term management of DMSMS in systems requires addressing the problem on three different management levels: reactive, pro-active and strategic.

7.5.1.1 *Reactive mitigation of obsolescence*

The reactive management level is concerned with determining an appropriate, immediate resolution to the problem of components becoming obsolete, executing the resolution process and documenting/tracking the actions taken. Many mitigation strategies exist for reactively managing obsolescence once it occurs. Replacement of parts with non-obsolete substitute or alternative parts can be done as long as the burden of system re-qualification is not unreasonable. There are also aftermarket electronic part sources, ranging from original manufacturer-authorized aftermarket sources that fill part needs with a mixture of stored devices (manufactured by the original manufacturer) and new fabrication in original manufacturer-qualified facilities. The secondary market ranges from original-manufacturer approved aftermarket suppliers to non-approved brokers and even eBay. However, buying obsolete parts on the secondary market from non-authorized sources carries its own set of risks – namely, the possibility of counterfeit parts (see Section 7.5.2). SRI International operates GEM and AME, which are electronic part emulation foundries that fabricate obsolete parts that meet original part qualification standards using newer technologies (i.e., BiCMOS gate arrays). Thermal uprating of commercial parts to meet the extended temperature range requirements of an obsolete Mil-Spec part is also a possible obsolescence mitigation approach.

7.5.1.2 *Lifetime buys*

Lifetime buy (also known as: "life-of-need buy", "all-time buy") is one of the most prevalent reactive obsolescence mitigation approaches

---

are characterized as slow clockspeed industries. In addition, because of the expensive nature of sustainment-dominated products (e.g., power plants, airplanes and ships) customers can't afford to replace these products with newer versions very often (slow clockspeed customers). DMSMS-type obsolescence occurs when slow clockspeed industries must depend on a supply chain that is organized to support fast clockspeed products.

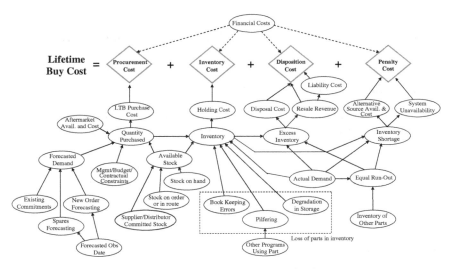

Fig. 7.4. Lifetime buy costs [7.6].

employed for DMSMS management. Lifetime buy means purchasing sufficient parts to meet all current and future demands. This is simpler in theory than in practice, due to many interacting influences and the complexity of multiple concurrent buys, as shown in Figure 7.4. The lifetime buy problem fundamentally has two facets: demand forecasting, and optimizing the buy quantities based on the demands forecasted.

Forecasted demand depends on sales forecasts and sustainment expectations (spares) for fielded systems (sparing is addressed in Chapter 4). The second aspect of the problem is determining how many parts should be purchased (the lifetime buy quantity).

Given a demand forecast, the quantities of parts necessary to minimize life-cycle cost can be calculated (depending on the penalty for running short or running long on parts, these quantities could be different than what simple demand forecasting predicts). In general, this is an asymmetric problem, where the penalty for underbuying parts and overbuying parts are not the same; if they were the same, then the optimum quantity to purchase would be exactly the forecasted demand. For example, the penalty for underbuying parts is the cost to acquire additional parts long after they become obsolete or, worse case, designing the part out of the system; the penalty for overbuying parts is paying for extra parts and for the holding (inventory or storage) cost of those parts for a long period and the possible loss of the investment in the parts.[7] In general, for sustainment-dominated systems, the penalty for underbuying parts is significantly greater than the penalty for overbuying parts.

Lifetime buy optimization is more generally referred to as the final-order problem, which is a special case of the newsvendor problem[8] from traditional operations management. Existing final-order models are intended for systems like manufacturing machinery that have long-term service contracts. To be able to provide long-term service, a manufacturer must be able to supply parts throughout the service period. However, the duration of the service period is typically much longer than the period over which the machine is manufactured. The period after the machine has been taken out of manufacturing is called the end-of-life service period (EOL). To avoid out-of-stock situations during the EOL, an initial stock of spare parts is ordered at the beginning of the EOL. This initial stock is called the final order.

The factors relevant to solving this problem are:

C_O = the overstock cost – the effective cost of ordering one more unit than what you would have ordered if you knew the exact demand (i.e., the effective cost of one left-over unit that can't be used, sold or returned).

C_U = the understock cost – the effective cost of ordering one fewer unit than what you would have ordered if you knew the exact demand (i.e., the penalty associated with having one less unit than you need or the loss of one sale you can't make).

Q = the quantity ordered.

D = uncertain demand.

The newsvendor problem is a classic example of an optimal inventory problem. As an example, consider a newsvendor who purchases

---

[7]Additionally, you may need to pay to dispose of the extra parts. The cost of disposal could be negative (reselling the parts) or positive (ensuring that parts are destroyed so they won't become a liability).

[8]The newsvendor problem seeks to find the optimal inventory level for an asset, given an uncertain demand and unequal costs for overstock and understock. This problem dates back to an 1888 paper by Edgeworth [7.7].

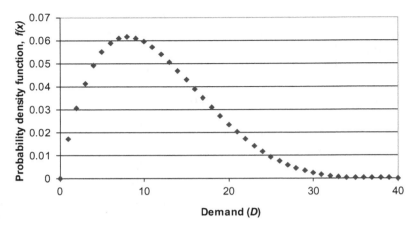

Fig. 7.5.   Demand forecast.

newspapers in advance for $0.20/paper. The papers can be sold for
$1.00/paper. The demand is represented by a beta distribution with
shape parameters: $\alpha = 2$ and $\beta = 5$ (lower bound 0, upper bound
40), which is shown in Figure 7.5.[9] How many papers should the
newsvendor buy in order to maximize their profit?

In this case $C_U = \$1.00 - \$0.20 = \$0.80$ ($0.80 is lost for each
sale that cannot be fulfilled) and $C_O = \$0.20$ ($0.20 is lost for each
paper purchased that cannot be sold).

Table 7.1 shows the calculations when $Q = 10$ (assuming discrete
demand). The quantities in Table 7.1 are determined using:

$$\text{Overstock Cost} = (Q - D)C_O, \text{ when } D \leq Q \qquad (7.1)$$

$$E[C_O] = f(x)(\text{Overstock Cost}) \qquad (7.2)$$

$$\text{Understock Cost} = (D - Q)C_U, \text{ when } D \geq Q \qquad (7.3)$$

$$E[C_U] = f(x)(\text{Understock Cost}) \qquad (7.4)$$

---

[9]The analysis presented here can be done with any distribution. A Beta distribution was chosen because it has a defined lower bound (i.e., it does not go to
$-\infty$).

Table 7.1. Newsvendor problem calculations for $Q = 10$.

Demand (D)	$f(x)$	Overstock Quantity (Q − D)	Overstock Cost	$E[C_0]$	Understock Quantity (D − Q)	Understock Cost	$E[C_U]$
0	0	10	2	0			
1	0.0169441	9	1.8	0.030499			
2	0.030544	8	1.6	0.04887			
3	0.0411803	7	1.4	0.057652			
4	0.0492075	6	1.2	0.059049			
5	0.0549545	5	1	0.054955			
6	0.0587257	4	0.8	0.046981			
7	0.0608016	3	0.6	0.036481			
8	0.06144	2	0.4	0.024576			
9	0.0608766	1	0.2	0.012175			
10	0.0593262	0	0	0	0	0	0
11	0.0569831				1	0.8	0.045586
12	0.0540225				2	1.6	0.086436
13	0.0506011				3	2.4	0.121443
14	0.0468579				4	3.2	0.149945
15	0.0429153				5	4	0.171661
16	0.03888				6	4.8	0.186624
17	0.0348435				7	5.6	0.195124
18	0.0308834				8	6.4	0.197654
19	0.027064				9	7.2	0.194861
20	0.0234375				10	8	0.1875
21	0.0200445				11	8.8	0.176392
22	0.0169151				12	9.6	0.162385
23	0.0140697				13	10.4	0.146325
24	0.01152				14	11.2	0.129024
25	0.0092697				15	12	0.111237
26	0.0073155				16	12.8	0.093639
27	0.005648				17	13.6	0.076813
28	0.0042525				18	14.4	0.061236
29	0.0031098				19	15.2	0.047269
30	0.0021973				20	16	0.035156
31	0.0014897				21	16.8	0.025027
32	0.00096				22	17.6	0.016896
33	0.0005803				23	18.4	0.010678
34	0.0003227				24	19.2	0.006197
35	0.0001602				25	20	0.003204
36	0.0000675				26	20.8	0.001404
37	2.195E-05				27	21.6	0.000474
38	4.453E-06				28	22.4	9.98E-05
39	2.856E-07				29	23.2	6.63E-06
40	0				30	24	0

The total expected loss in this case is given by

$$\text{Expected Total Loss} = \sum_{i=0}^{40} \text{E}[C_{O_i}] + \sum_{i=0}^{40} \text{E}[C_{U_i}]$$

$$= \$0.37 + \$2.64 = \$3.01 \qquad (7.5)$$

The result in Equation (7.5) means that if the newsvendor purchases $Q = 10$ newspapers, he can expect to lose $\$3.01$.[10] If the analysis in Table 7.1 is repeated for $Q = 16$, the total loss $= \$1.97$, which indicates that buying 16 newspapers instead of 10 is better (a smaller loss). So, what is the value of Q that minimizes the expected total loss – that is, what is the optimum number of newspapers for the newsvendor to purchase?

If we let the expected total loss as a function of Q be denoted by $L(Q)$, and assume a continuous demand, then

$$L(Q) = C_O \int_0^Q (Q - x) f(x) dx + C_U \int_Q^\infty (x - Q) f(x) dx \qquad (7.6)$$

Equation (7.6) expresses exactly what was shown discretely in Table 7.1 and Equation (7.5), where $f(x)$ is the probability density function of the demand. The first term in Equations (7.5) and (7.6) is the expected cost of overstocking (having too many) and the second term is the expected cost of understocking (having too few). Taking the derivative of both sides of Equation (7.6) and setting it equal to zero to find a minimum gives,

$$\frac{dL(Q)}{dQ} = C_O F(Q) - C_U[1 - F(Q)] = 0 \qquad (7.7)$$

where $F(Q)$ is the cumulative distribution function of the demand (the in-stock probability),

$$F(Q) = \int_0^Q f(x) dx \qquad (7.8)$$

---

[10] Depending on the type of demand distribution used, the sums in Equation (7.5) may actually go to ∞. In this example, a beta distribution with a fixed upper bound of 40 was used so the sums are complete for this example (i.e., no terms are omitted).

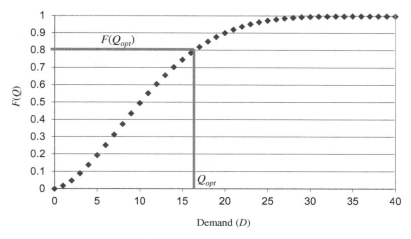

Fig. 7.6. $F(Q)$ versus demand (D).

The value of Q that satisfies Equation (7.7) is denoted by Q_{opt}, which is defined by

$$F(Q_{opt}) = \frac{C_U}{C_O + C_U} \qquad (7.9)$$

Equation (7.8) is called the critical ratio (or critical fractile) and is valid for any demand distribution (any $f(x)$). At Q_{opt}, the marginal cost of overstock is equal to the marginal cost of understock. At Q_{opt}, $F(Q_{opt}) = \Pr(D \leq Q_{opt})$.

For the example given earlier in this section, $F(Q)$ is shown in Figure 7.6 and $F(Q_{opt}) = 0.8$ from Equation (7.9), which corresponds to $Q_{opt} = 16.9$ in Figure 7.6.

The solution discussed in this section assumes that backlogs are not allowed (i.e., unfulfilled demand is lost) and carryover is not allowed (i.e., leftover inventory has zero salvage value).

How can the newsvendor problem analysis be applied to lifetime-buying spare parts? Consider the bus problem introduced in Chapter 4 (Section 4.4.1.2 and Section C.2.3) where the bus is intended to operate for 200,000 miles per year. A critical failure mode has been identified for a component and reliability analysis indicates that the failures have a constant failure rate with $\lambda = 1.4 \times 10^{-5}$ failures/mile. Assume there will be no future opportunity to procure additional spare parts. How many spares should be bought per bus now to support 10 years worth of its operation? Assume that the part costs $2000 to procure now. Assume if you run out of parts you will have to procure the part from a third party at an effective cost of $20,000/part. The number of renewals for a 10-year period was computed in Section 4.4.1.3, it is 2.697/year. So, should we buy $(10)(2.697) = 26.97$ spares? If the cost penalties were symmetric, this would be the right buy size, but the penalties are asymmetric. In this case $C_U = \$20,000 - \$2000 = \$18,000$, $C_O = \$2000$, so from Equation (7.9) $F(Q_{opt}) = 0.9$, this is the confidence level for the demand distribution. Where do we get the demand distribution? Actually, the demand distribution is generated using a discrete-event simulator in Section C.2.3. In Figure C.2, the mean is 2.697, however 2.788/year are necessary to satisfy 90% confidence, so $(10)(2.788) = 27.88$ parts need to be purchased at the lifetime buy per bus to minimize the life-cycle cost.

In this simple treatment, there is an important implicit assumption and several key elements of the problem have been omitted. A "must support" assumption is implicit in lifetime buy problems that can significantly increase the magnitude of the penalty associated with running out of parts. In the example above, you cannot choose not to support the product – that is, you are not allowed to fail to fulfill the demand then you must pay the penalty to purchase extra parts from the broker if you run out. Another significant assumption that is implicitly made in the classical newsvendor problem is that there is no time dependence. For the newsvendor, the time periods between purchasing newspapers and selling, or running out of papers are short. For the lifetime buy problem, this is not true. For lifetime buys of parts to support sustainment-dominated systems, the parts are purchased, placed in inventory, and drawn from inventory over years, and if you run short of parts, the penalty is assessed near the end of the support period many years after the lifetime buy was made. In this case the cost of money (non-zero discount rate) and the cost of holding parts in inventory will play significant (potentially dominate) roles. The part lifetime buy problem is analogous to the newspaper boy buying an inventory of papers in year 2010, paying to store the papers as they are gradually sold over a 10-year period, and then either having extra papers that can't be sold or customers that can't be satisfied in year 2025.

The inclusion of the cost of money in the discrete newsvendor problem solution does not affect the $E[C_O]$ term in Equation (7.5) because the overbuy occurs at the beginning of the analysis (the beginning of year 1) if money is in year 0 dollars. However, the $E[C_U]$ term is impacted because the penalty for underbuying occurs after the order quantity, Q, runs out, which is (presumably) near the end of the demand.

Although there are many extensions to the classical newsvendor formulation that accommodate a variety of different situations, detailed discrete-event simulators are generally used to model time-dependent cost of money and holding cost effects, see Problem 7.4.

7.5.1.3 *Design refresh planning (DRP)*

Lifetime buy discussed in the previous section is a *reactive* approach to obsolescence management—that is, minimizing the cost of resolving the problem after it has occurred. While reactive solutions always play a major role in obsolescence management, ultimately, higher payoff is possible through *strategic* management approaches.

Because of the long manufacturing and field life associated with sustainment-dominated systems, they may be refreshed or redesigned one or more times during their life to update functionality and manage obsolescence. Unlike high-volume commercial products for which redesign is driven by improvements in manufacturing, equipment or technology, for sustainment-dominated systems, design refresh[11] is often driven by obsolescence that would otherwise render the product un-producible and/or un-sustainable.

A methodology that determines the best dates for design refreshes and the optimum reactive management approaches to use between the refreshes is needed.

The simplest model for performing life-cycle planning associated with technology obsolescence (specifically, electronic part obsolescence) was developed by Porter [7.8]. Porter's approach focuses on

---

[11] Refresh refers to changes that "have to be done" in order for the system functionality to remain usable. Redesign or technology insertion implies "want to be done" system changes, which include adopting new technologies to accommodate system functional growth and/or to replace and improve the existing functionality of the system.

calculating the present value (PV) of last-time (bridge) buys[12] and design refreshes as a function of the design refresh date. As a design refresh is delayed, its PV decreases and the quantity (and thus, cost) of parts that must be purchased in the last-time buy required to sustain the system until the design refresh takes place increases. Alternatively, if design refresh is scheduled relatively early, then last-time buy cost is lower, but the PV of the design refresh is higher. In a Porter model, the cost of the last-time buy (C_{LTB}) is given by

$$
C_{LTB} = \begin{cases} 0 \text{ when } i = 0 \text{ or if } Y_R = 0 \\ P_0 \sum_{i=1}^{Y_R} Q_i + C_h \sum_{i=1}^{Y_R} \left[\left(\sum_{k=i}^{Y_R} Q_k - \frac{Q_i}{2} \right) / (1+r)^i \right] \\ \quad \text{if } Y_R > 0 \end{cases}
$$

$$(7.10)$$

where

i is the year,

P_0 is the price of the obsolete part in the year of the last-time buy (beginning of year 1 in this case),

Y_R is the year of the design refresh ($0 =$ year of the last-time buy, $1 =$ one year after the last time buy, etc.), Q_i is the number of parts needed in year I,

C_h is holding cost per part per year, and

r is the discount rate (or WACC).

Equation (7.10) assumes that the part becomes obsolete at the beginning of year 1 and that the last-time buy is made at the beginning of year 1. The first term in Equation (7.10) is the cost of buying the last-time buy of parts and the second term is the cost of holding (storing) the parts until they are needed. The second term assumes that all the holding cost for a year are paid at the end of the year and that there is a constant demand for parts throughout the year.[13]

---

[12] A last-time or bridge buy means buying a sufficient number of parts to last until the part can be designed out of the system at a design refresh. Last-time buys become lifetime buys when there are no more planned refreshes of the system.

[13] The original Porter model [7.8] omits the holding cost for the last-time buys, i.e., assumes that $C_h = 0$.

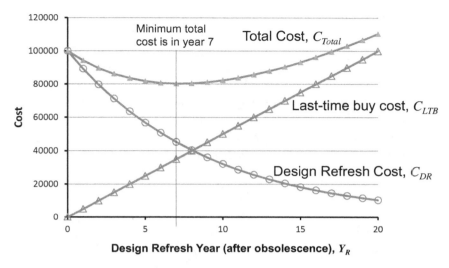

Fig. 7.7. Example application of Porter's design refresh costing model.

The design refresh cost for a refresh in year Y_R (in year 0 dollars), C_{DR}, is given by

$$C_{DR} = \frac{C_{DR_0}}{(1+r)^{Y_R}} \tag{7.11}$$

where C_{DR_0} is the design refresh cost in year 0.

The total cost for managing the obsolescence with a year Y_R design refresh is given by

$$C_{Total} = C_{LTB} + C_{DR} \tag{7.12}$$

Figure 7.7 shows a simple example using the Porter model. In this case $C_{DR_0} = \$100,000$, $r = 12\%/\text{year}$, $Q_i = 500$ (for all i from year 1 to 20, $Q_i = 0$ thereafter), $C_h = 0$, and $P_0 = \$10$. In this simple example, the model predicts that the optimum design refresh point is in year 7.

The optimum refresh year from the Porter model can be solved for directly for a simplified case. Substituting Equations (7.10) and (7.11) into Equation (7.12) and assuming that the holding cost (C_h)

is 0 and that the demand quantity is the same in every year, we get,

$$C_{Total} = P_0 \sum_{i=1}^{Y_R} Q_i + \frac{C_{DR_0}}{(1+r)^{Y_R}} = P_0 Q Y_R + C_{DR_0} e^{-rY_R} \qquad (7.13)$$

Equation (7.13) assumes that $Q = Q_i$ for all $i = 1$ to Y_R and that $1/(1+r)^{Y_R} \cong e^{-rY_R}$ (which is a continuous compounding assumption). The minimum value of C_{Total} can be found by setting the derivative of Equation (7.13) with respect to Y_R equal to zero:

$$\frac{dC_{Total}}{dY_R} = P_0 Q - rC_{DR_0} e^{-rY_R} = 0 \qquad (7.14)$$

Solving Equation (7.14) for Y_R we get

$$Y_R = \frac{1}{-r} \ln \left(\frac{P_0 Q}{rC_{DR_0}} \right) \qquad (7.15)$$

Equations (7.14) and (7.15) are only applicable when $r > 0$ (non-zero discount rate) and $rC_{DR_0} \geq P_0 Q$. For cases where $r = 0$ or $rC_{DR_0} < P_0 Q$ the optimum design refresh date is at $Y_R = 0$. It should be pointed out that the Y_R appearing in Equations (7.13)–(7.15) is the Y_R that minimizes life-cycle cost, whereas the Y_R appearing in Equations (7.10) and (7.11) is a selected refresh year. For the example shown in Figure 7.7, Equation (7.15) gives $Y_R = 7.3$ years.

At its simplest level, the conceptual basis for the construction of the basic Porter model is similar to the construction of EOQ (Economic Order Quantity) models (see Section 6.2.1). In the case of EOQ models, the sum of the part cost (purchase price and holding/carrying cost) and the order cost is minimized to determine the optimum quantity per order. The Porter model has a similar construction where the part cost is the same as in the EOQ model (with the addition of the cost of money) and the order cost is replaced by the cost of design refreshing the system to remove the obsolete part.

The Porter model has limited applicability for several reasons. First, the Porter model only treats the cost of supporting the system up to the design refresh, i.e., there is no accommodation for costs incurred after the design refresh. In the Porter model, the analysis terminates at Y_R. This means that the time span between the refresh (Y_R) and the end of support of the system is not considered, i.e., the

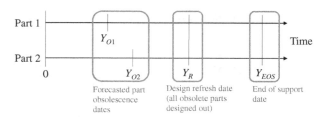

Fig. 7.8. Multi-part, single design refresh timeline.

costs associated with buying parts after the design refresh to support the system to some future end-of-support date are not included in the determination of the optimum design refresh date. Second, the Porter model only considers solutions with a single refresh in the system's life. In order to treat multiple design refreshes in a product's lifetime, Porter's analysis could be reapplied after a design refresh to predict the next design refresh, effectively optimizing each individual refresh, but the coupled effects of multiple design refreshes (coupling of decisions about multiple parts and coupling of multiple refreshes) in the lifetime of a product are not accounted for, which is a significant limitation for the application of the Porter approach to real systems.

A multi-part Porter model (still solving for a single refresh) can be formulated considering the timelines in Figure 7.8. The total cost of managing the timeline in Figure 7.8 is given by

$$C_{Total} = \sum_{i=1}^{N} C_{before_i} + \sum_{i=1}^{N} C_{LTB_i} + \sum_{i=1}^{N} C_{H_i} + C_{DR_0} + \sum_{i=1}^{N} C_{after_i}$$

$$(7.16)$$

Where the first term is the cost of buying parts as needed from time 0 to their respective obsolescence dates. The second term is the cost of last-time buys of the parts at their obsolescence dates. The third term is the cost last-time buy holding cost. The fourth term is the design refresh cost (all obsolete parts addressed). The last term is the cost of buying parts as needed from Y_R to Y_{EOS}. The original Porter model only includes the second and fourth costs for a single part ($N = 1$).

Equation (7.16) can be expanded as

$$
\begin{aligned}
C_{Total} \cong & \sum_{i=1}^{N} \left[P_i Q_i \sum_{j=0}^{Y_{O_i}} \frac{1}{(1+r)^j} \right] + \sum_{i=1}^{N} \left[\frac{P_i Q_i (Y_R - Y_{O_i})}{(1+r)^{Y_{O_i}}} \right] \\
& + \sum_{i=1}^{N} \left[\frac{C_{h_i} Q_i}{2} \sum_{j=0}^{Y_R - Y_{O_i}} \frac{2j-1}{(1+r)^{Y_{O_i}+j}} \right] + \frac{C_{DR_o}}{(1+r)^{Y_R}} \\
& + \sum_{i=1}^{N} \left[P_i Q_i \sum_{j=Y_R}^{Y_{EOS}-1} \frac{1}{(1+r)^j} \right] \qquad (7.17)
\end{aligned}
$$

where C_{h_i} is the annual holding cost per part for the ith part, r is the discount rate (WACC), Q_i is the part demand for the ith part, and P_i is the purchase price per part for the ith part. The holding cost term assumes that parts are consumed at a constant rate.

Equation (7.17) can be solved for the value of Y_R that minimizes C_{Total}.

$$
\begin{aligned}
\frac{dC_{Total}}{dY_R} = & \sum_{i=1}^{N} \left[\frac{P_i Q_i}{(1+r)^{Y_{Oi}}} \right] + \sum_{i=1}^{N} [H_i Q_i (Y_R - Y_{Oi})] \\
& - \frac{C_{DR_o} \ln(1+r)}{(1+r)^{Y_R}} - \sum_{i=1}^{N} \left[\frac{P_i Q_i}{(1+r)^{(Y_R-0.5)}} \right] = 0 \quad (7.18)
\end{aligned}
$$

If parts only went obsolete at most one time during the sustainment life of a system, and only one refresh was allowed, then the value of Y_R found by solving Equation (7.18) would be the optimum refresh date. However, parts often go obsolete several times during a critical system's support life,[14] and it is possible that the optimum consists of multiple refreshes.

---

[14]The replacement for a part at a refresh will itself become obsolete before the end of the support life for systems. This is common for electronic parts.

7.5.1.4 *Non-hardware obsolescence*

Obsolescence affects more than hardware. It also impacts software, workforces, and intellectual property – all of these things are part of the supply chain too.

The applicable definition of software obsolescence varies depending on the system that uses the software, and where and how that system is being used. Commercial software has both end-of-sale dates and end-of-support dates that can be separated by long periods of time. For many mainstream commercial software applications (e.g., mass-market operating systems), both the end-of-sale and end-of-support dates may be published by the software vendors. For applications that have a connection to the public web (e.g., servers and communications systems), the relevant software obsolescence date for both the deployment of new systems and the continued use of fielded systems is often the end-of-support date, because it is the date on which security patches for the software terminate, making the continued use of the software a security risk that is unacceptable for many critical systems. For other embedded or isolated applications, the relevant software obsolescence date is governed by either an inability to obtain the necessary licenses to continue using it or changes to the system that embeds it (functional obsolescence issues). See [7.9] for more discussion of software obsolescence.

Obsolescence isn't confined to just hardware and software. Many types of systems that have to be supported for long periods of time lose critical portions of their workforce before the support for the system ends. The loss of critical workforce does not refer to the normal turnover of unskilled labor, but rather, the loss of highly-skilled workers that have unique experience and are either non-replenishable or would take very long periods of time to reconstitute. There is lots of existing research on "skills obsolescence", i.e., people who have obsolete skills and need to be retrained in order to be employable. The type of obsolescence referred to here is the opposite of skill obsolescence, it is "critical skills loss", e.g., [7.10]. See Section 8.1 for more discussion of workforce obsolescence.

The phrase "data obsolescence" has been used in the context of data "aging out", i.e., data that no longer complies with accepted data models/standards or data that is simply no longer accessible.

More generally, data obsolescence is sometimes used to imply "digital obsolescence" that occurs when a digital resource can no longer be accessed because of its obsolete format, e.g., data on magnetic tape or a 3.5 inch floppy disk. Digital obsolescence can be due to physical media, the hardware (required to read the media), or the availability of the software that runs on the hardware. We are starting to talk about "data obsolescence" in the context of using additive manufacturing for sustainment applications – this is digital obsolescence, but has a bit broader meaning. One of the overriding concerns for AM is data provenance, i.e., has the data that defines the part you need to print been corrupted, compromised, or become obsolete, where obsolescence means that the data is no longer consistent with the printer, the materials, the environment in which you have to print the part, or the tests you need to run to verify the finished part. See Section 10.1 for more discussion of AM data issues.

7.5.2   *Counterfeit parts*

A counterfeit is a fraudulent imitation of something.[15] For parts, the definition of a counterfeit is a part that is misrepresented as something that it is not. The misrepresentation could be the type of part (one part number misrepresented as a different part number), the quality of the part, the origin of the part, or a new versus used part. The obsolescence of parts discussed in Section 7.5.1 can precipitate the appearance of counterfeit versions of the obsolete part (although not all counterfeit parts are necessarily obsolete).

In this section we are narrowing the counterfeiting scope to hardware in critical systems, where system failures (or incorrect operation) due to counterfeit parts directly or indirectly results in risks to human life, national security, and significant economic peril.

---

[15]In the context of supply-chain risks for systems, we refer to "fake" parts as counterfeits rather than forgeries. Counterfeiting is making or creating an unauthorized imitation of a genuine article (generally en masse) with the intent to defraud. Usually the word forgery is used to describe the unauthorized reproduction of valuable unique objects where the value is determined not by the actual object but the social or mutual agreement or culture (forgery is applied to art and historical documents). Forgery is a crime, while counterfeiting is criminal imitation. Forgery is about the act where counterfeit is about the result.

Counterfeiting: Not a New Problem

Counterfeit products and intellectual property piracy have existed since the beginning of commerce. While at this time, the United States views itself as a victim of IP theft in the global arena, when the U.S. was a fledgling nation it, in fact, often actively "borrowed" high technology from the developed nations of the time, particularly the United Kingdom. In his "Report on Manufactures", Treasury Secretary Alexander Hamilton advocated bringing "improvements and secrets of extraordinary value" into the country and developing a method of rewarding them [7.11]. The reasons that prompted the U.S. to "borrow" IP in the 1700s is one of the reasons why counterfeiting exists today, i.e., trade barriers and restrictions. In the 1790s, the British government banned the export of textile machinery and the emigration of workers who operated the machines in an attempt to create global dependency on them [7.12, 7.13].

For example, [7.14] provides a list of airline incidents that resulted in loss of life and loss of the aircraft, which have been attributed to hardware counterfeits. Defense systems are also vulnerable, in 2006, fifty Terminal High Altitude Area Defense (THAAD) missiles had counterfeit flash memory chips, which stored critical flight software, installed on their mission computer. Failure of these chips would likely lead to failure of the missile itself [7.15]. Electronic parts and hardware counterfeiting has a disproportionate impact on the critical systems.

There are many motivations for counterfeiters, which range from industrial espionage to market share increase, but most are simply rooted in supply and demand. In the field of electronics and particularly for electronic parts, the majority of reported instances of counterfeiting relate to parts that are obsolete, but are still needed to maintain old systems. There is also a large customer segment that encounters counterfeiting that exists due to unpredictable shortages (of non-obsolete parts) often caused by natural disasters, political upheavals, and even pandemics – these are referred to as allocation problems (see Section 7.5.3). The counterfeiters are agile, they take advantage of the situation and target the specific products impacted by the supply-chain disruptions caused by the situations described above, but the customer's logistics and technology are often ill-prepared to avoid and detect such problems.

In the field of electronic parts, it is often assumed that if a component is purchased from an authorized source then the risk of obtaining counterfeit is low; in fact, most counterfeits find their way to customers through brokers and independent distributors. However, the reality is far more complex. There have been incidents of reports of counterfeit part sales by a range of entities including: brokers, independent distributors, authorized distributors, contract manufacturers, OEMs, government prime/subcontractors, U.S. DoD depots, OCMs, U.S. federal agencies, the Defense Logistics Agency, and U.S. state/local government, Figure 7.9.[16]

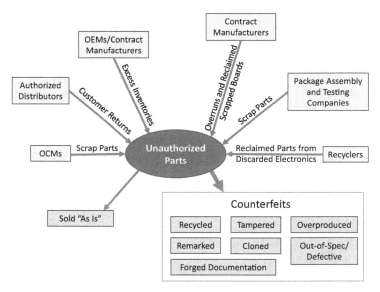

Fig. 7.9.   Common sources of counterfeit electronic parts [7.16].

---

[16]Note, "product diversion" does not appear in Figure 7.9, but is relevant for many critical systems (e.g., medical). Diversion occurs when a product developed and legally sold into a specific market, e.g., with materials that are authorized

Critical systems, including medical, aerospace, transportation, infrastructure, energy generation and defense have several features in common making them particularly vulnerable to counterfeit parts – their production volumes are low, their systems must be supported for very long periods of time, and the cost of design changes and approvals (qualification and certification) are often prohibitive. All of these attributes taken together results in these systems regularly seeking parts from non-traditional channels and thereby increasing the risk of getting counterfeit parts. The electronic resellers association international (ERAI) found that reported counterfeits of "military" temperature range integrated circuits (used in aerospace, defense, energy generation, industrial controls, and a host of other applications) is consistently an order of magnitude higher than estimated military system usage and these parts are more counterfeited than other temperature ranges [7.17].

The impacts of counterfeit parts (electronics in particular) go well beyond just breach of contract and low-quality products. Additional threats and security challenges related to the supply chain include IP piracy, trojan insertion, and reverse engineering.[17] While concerns are often raised about the intervention of nation states by tampering with high-complexity parts that are inserted into critical systems (e.g., energy generation, infrastructure), the occurrences of such instances are extremely rare and their detection and avoidance are the focus of intelligence operations not the disruption of commercial illicit supply chain operations.

7.5.3 *Product allocation*

Product allocation is "an amount or portion of a resource assigned to a particular recipient". In practice, allocation is the process (algorithm) by which a part supplier allocates part stocks to their customers when there is a shortfall of parts, i.e., when there are not

---

for the intended market, are repackaged and sent into a market (e.g., a country) where the materials, validation, qualification, etc., are not authorized.

[17]A particularly insidious form of compromise is a parametric or process trojan, i.e., physical features within an integrated circuit are modified (e.g., trace widths, oxide thicknesses), which could result in a reduction in reliability.

enough parts available to fulfill all the demand. Allocation (shortage) issues arise when the supply chain cannot satisfy demand. The supply chain's inability to satisfy demand may be due to many different reasons.

A Perfect Storm

In 2020, an explosion in the demand for consumer electronics (smart phones and PCs) driven in part by the COVID-19 pandemic caused a global supply shortage of some electronic parts. By early 2021, the average lead times for chips jumped to over 15 weeks. Pat Gelsinger, CEO of Intel Corporation, said the semiconductor shortage could potentially stretch into 2023 [7.18].

The electronics supply chain was already disrupted before the pandemic by a tariff war between the United States and China that was forcing the relocation of several significant electronics manufacturers from China to Southeast Asia. The existing disruptions were exacerbated by uncertainties in supply due to the COVID-19 pandemic resulted in sharp swings in orders in 2020, which adds confusion to the market as chip fabricators try to match capacity with demand. The shortage market caught the automotive inventory planners by surprise causing some automakers to suspend production in 2021 because of shortages, while making PlayStations® and Xboxes® harder to find in stores.

In the supply chain for critical systems, allocation problems can arise for components that while not obsolete, nonetheless have extremely long delivery times (e.g., months and possibly years). This can be caused by natural disasters, political unrest, pandemics, etc., which limit the quantity of components available to the market. When demand significantly exceeds supply, usually the largest customers (e.g., highest-volume customers) are supplied before low-volume customers, which often results in critical systems customers going to the "back of the line" for their components.

Additional risks are indirectly related to allocation – purchase from unauthorized sources and related risk of obtaining lower quality or counterfeit parts (see Section 7.5.2). There is a general belief that shortage markets accelerate product obsolescence and creates a market for counterfeit parts. Critical system manufactures and sustainers should not expect any special treatment from a shortage market.

Some critical systems may be covered by the Defense Production Act of 1950 (DPA) and thereby can be given allocation priority under some circumstances. The DPA was part of a broad civil defense and war mobilization effort during the Cold War. The DPA provides the President of the United States with the authorization to:

1. Require businesses to accept and prioritize contracts for materials deemed necessary for national defense, regardless of losses they may incurred in their business. The DPA also allows the President to prohibit hoarding or price gouging for designated materials.
2. Establish mechanisms (such as regulations, orders or agencies) to allocate materials, services and facilities to promote national defense.
3. Control the civilian economy so that scarce and critical materials necessary to the national defense effort are available for defense needs.

7.5.4 *Inventory aging*

An organization may mitigate (or believe they are mitigating) aging supply-chain problems by procuring and storing enough parts to fulfill their anticipated needs. Some types of parts can be stored indefinitely, but not all types of parts. Depending on the type of part, there may be a finite "shelf life" after which the part must be retested, or may simply have to be disposed of (shelf life is discussed in Section 6.6).

7.6 Supply-Chain Trust

The "Gotta Buy a Thing from a Guy ..." example (at the end of Chapter 1) fundamentally relies on developing and maintaining supply-chain trust in order to work. In general, the components of trust in supply chains are: honesty, loyalty, fairness, openness and competence, and trust only exists when both parties think it exists. In the context of critical systems, trust is defined as the confidence in one's ability to secure systems by assessing the integrity of the people and processes used to design, generate, manufacture and distribute those systems and the components they are composed of.

Ensuring trust in the supply chain for critical systems has evolved along with the supply chain. One approach to ensuring trust for U.S. government systems is through Federal Acquisition Regulations (FARs) (see Section 9.3) that are placed on tier 1 suppliers and flowed by them down through their supply chains. Although the FARs may be effective in the top-most tiers of the supply chain they defuse quickly, since visibility and enforcement prove to be impossible in the lower tiers of the supply chain as a result of contract privity. One symptom of this diffusion is that not all the nodes (the dots in Figure 7.1) are trusted suppliers, some are untrusted (the ones you simply don't know anything about), but it doesn't change the fact that they are in the supply chain.[18] CMMC [7.19] and other certifications (essentially a *Good House Keeping* seal of approval) flow down through the contracts, but are also fundamentally unenforceable.

The U.S. DoD adopted the concept of "trusted foundries" in the 1990s (this concept was developed and applied to microelectronics). In the trusted foundry model, security and risk are controlled by constraining procurements to a supply chain that is limited and trusted. Within this context, "trusted sources" are assumed to:

- Provide an assured a "Chain of Custody" for components
- Ensure that there will not be any reasonable threats related to disruption in supply
- Prevent intentional or unintentional modification or tampering with components
- Protect the components from unauthorized attempts at reverse engineering, exposure of functionality or evaluation of their possible vulnerabilities.

---

[18] Supply chains with untrusted players are the norm, not the exception. If you are able to look deep enough into every supply chain you will find untrusted sources.

The biggest problem with trusted foundries is that relatively few foundries want to participate, and critical system manufacturers want (need) access to a broader set of vendors and technologies. It is also unclear if trusted foundries can really deliver the security promised as they may also have deep supply chains and their ability to guarantee trust is unclear for all the reasons already articulated.

In 2020, the DoD announced a shift away from trusted foundries to zero-trust approaches [7.20]. Zero-trust architectures (ZTA) accepts the inevitability that supply chains have flaws and seeks to find a mix of system architecture, system health management (e.g., anomaly detection), supply-chain illumination (e.g., vetting), and risk management that creates and maintains safe and usable systems from supply chains with minimal trust [7.21]. ZTA has been developed as an organizational security model [7.22], but not applied to supply chains.

7.6.1 *Blockchain*

Blockchain technology is a potential trust enabler for the supply chain security and risk mitigation. Blockchain allows greater visibility and improved collaboration between part origination and use. Blockchain may be a solution (or part of a solution) for disrupting counterfeit part supply chains, but an understanding of the sociotechnical network is needed to determine what must be required in order for it to work (and how it can be knowingly or unknowingly compromised).

Essentially, a blockchain is a data structure composed of a decentralized distributed digital ledger of data hosted among groups of independent participants. Theoretically a blockchain enables near real-time concurrent access, validation, and recording by multiple, decentralized participants. Blockchains use cryptographic techniques to generate digital fingerprints and signatures for the data transactions. Data integrity is preserved using consensus protocols that ensure that all nodes are synchronized, and digital signatures that authenticate the content. Ledgers cannot be changed without consensus (agreement), among all network participants (i.e., they are immutable).

Distributed Ledgers

Ledgers have been an important part of commerce for thousands of years. Ledgers are used as a record keeper for the exchange of goods, money and property.

A distributed ledger, is the record keeping for an asset that is shared across multiple sites, geographies, and/or institutions, which has no central administrator. All the participants have their own identical copy of the ledger and any changes to the ledger are reflected across all the entities. The Roman Empire's banking system is an early example of the application of a distributed ledger.

Blockchain is a subset of distributed ledgers, and it has additional functionality aside from the traditional distributed ledger's scope. Blockchains are usually comprised of blocks of data, whereas a distributed ledger is just a database. In a blockchain, the blocks are in a particular sequence and usually required a "proof of work" consensus.

A supply chain that is neither secured nor trusted opens up the opportunity for compromises, i.e., hardware (and software) produced from an untrusted supply chain cannot serve as the underlying root of trust. Therefore, the viability of using blockchain for the supply chain of critical parts is being explored [7.23]. Today, several electronics fabricators have implemented internal traceability into the components they manufacture. The primary reason for internal traceability is for quality control, e.g., so they can identify the specific wafer, wafer location, etc. that the part came from, e.g., see IPC-1783 [7.24]. This sort of traceability provides the relationship of materials to the product, i.e., when you find a problem, you are able to go back and find the source of the problem, however, it is not blockchain. Blockchain is not traceability, but blockchain can be used for traceability.

A fundamental concern with blockchain for critical system's supply chain is uncertainty about how the blockchain ages, i.e., a blockchain may function as designed today and in the near future for ensuring trust, but will the blockchain for a part manufactured today still be verifiable 20 years from today after the blockchain's participants have been through in various mergers and acquisitions, the part was long ago discontinued, and the participants in the original transactions for the part have moved on to different business areas? Will the original manufacturer and original authorized distributor of a part today still be willing (and able) to provide consensus for the part's blockchain 20 years from now? Blockchain technology can also be used to implement smart contracts; contracts that are executed automatically when certain conditions are met. These can be used to securely automate some supply chain functions.

References

[7.1] DuHadway, S., Carnovale, S., and Hazen, B. (2019). Understanding risk management for intentional supply chain disruptions: Risk detection, risk mitigation, and risk recovery, *Applications of OR in Disaster Relief Operations*, 283, pp. 179–198.

[7.2] Khan, O. and Burns, B. (2007). Risk and supply chain management: creating a research agenda, *International Journal of Logistics Management*, 18(2), pp. 197–216.

[7.3] Sandborn, P. (2008). Trapped on technology's trailing edge, *IEEE Spectrum*, 45(1), pp. 42–45.

[7.4] Song, Y. and Lau, H. (2004). A periodic review inventory model with application to the continuous review obsolescence problem, *European Journal of Operations Research*, 159(1), pp. 110–120.

[7.5] Fine, C. (1998). *Clockspeed: Winning Industry Control in the Age of Temporary Advantage* (Perseus Books, USA).

[7.6] Feng, D., Singh, P., and Sandborn, P. (2007). Optimizing lifetime buys to minimize lifecycle cost, *Proceedings of the 2007 Aging Aircraft Conference*.

[7.7] Edgeworth, F. (1888). The mathematical theory of banking, *Journal of Royal Statistical Society*, 51, pp. 113–127.

[7.8] Porter, G. Z. (1998). An economic method for evaluating electronic component obsolescence solutions, Boeing Company White Paper.

[7.9] Sandborn, P. (2007). Software obsolescence – Complicating the part and technology obsolescence management problem, *IEEE Transactions on Components and Packaging Technologies*, 30(4), pp. 886–888.

[7.10] Sandborn, P. A. and Prabhakar, V. J. (2015). The forecasting and impact of the loss of the critical human skills necessary for supporting legacy systems, *IEEE Transactions on Engineering Management*, 62(3), pp. 361–371.

[7.11] Ben-Atar, D. S. (2004). *Trade Secrets: Intellectual Piracy and the Origins of American Industrial Power* (Yale University Press, USA).

[7.12] Row, W. (1796). *Look Before You Leap*, 3rd Ed. (British Pamphlet).

[7.13] Jeremy, D. J. (1973). British textile technology transmission to the United States: The Philadelphia region experience, 1770–1820, *Business History Review*, 47(1), pp. 24–52.

[7.14] Wald, M. L. (1995). Counterfeit airliner parts are aid to be often used, *New York Times*, Section A, p. 18, May 25.

[7.15] Senate Armed Services Committee (2012). *Inquiry into Counterfeit Electronic Parts in the Department of Defense Supply Chain*, U.S. Government. Available at: https://www.armed-services.senate.gov/imo/media/doc/Counterfeit-Electronic-Parts.pdf.

[7.16] Das, D. (2021). Center for Advanced Life Cycle Engineering (CALCE), University of Maryland, personal communication.

[7.17] Schipp, F. (2018). Current trends in counterfeit electronic parts, *Proceedings of the CALCE Symposium on Counterfeit Parts and Materials*, College Park, MD.

[7.18] Fitch, A. (2021). Intel CEO says chip shortage could stretch into 2023. *The Wall Street Journal*, July 22, Available at: https://www.wsj.com/articles/intel-intc-2q-earnings-report-2021-11626899296.

[7.19] CMMC, Version 1, Cybersecurity Maturity Model Certification, Carnegie Mellon University and The Johns Hopkins, University Applied Physics Laboratory LLC, January 30, 2020.

[7.20] Lopez, C. T. (2020). DoD Adopts "Zero Trust" Approach to Buying Microelectronics, *DoD News*, May 19. https://www.defense.gov/Explore/News/Article/Article/2192120/dod-adopts-zero-trust-approach-to-buying-microelectronics/.

[7.21] Kindervag, J. (2010). *No More Chewy Centers: Introducing the Zero Trust Model of Information Security*, Forrester Research.

[7.22] Rose, S., Borchert, O., Mitchell, S., and Connelly, S. (2020). *Zero Trust Architecture*, SP 800-207, NIST.

[7.23] Xu, X., Rahman, F., Shakya, B., Vassilev, A., Forte, D., and Tehranipoor, M. (2019). Electronics supply chain integrity enabled by blockchain, *ACM Transaction of Design Automation of Electronic Systems*, 24(3), pp. 1–25.

[7.24] IPC-1783, International Standard for Component-Level Authentication (CLA).

Problems

7.1 The system you are manufacturing requires 100 of a particular part per month. Your organization decides to put in place a 6-month buffer of parts to mitigate supply-chain disruptions (the alternative to the buffer is to purchase parts as needed). If the parts cost $14,000 each and the effective WACC is 0.9%/month, what is the cost of this buffering strategy, i.e., how much more does it cost than buying the parts as needed?

7.2 Assume you have to make a lifetime buy of an electronic part because it has become obsolete. Assume that all the future demand for the part (to continue manufacturing and supporting the product) is given by a beta distribution with $\alpha = 2$ and $\beta = 5$ (lower bound 900, upper bound 1200); the parts can be purchased for $2/part at the lifetime buy point. If the lifetime buy runs out the parts must be purchased from a broker for $30/part. What is the optimum number of parts to buy at the lifetime buy?

7.3 For the bus problem in Section 7.5.1.2, if the part costs $3000 to procure. Assume if you run out of parts you will have to procure the part from a third party at an effective cost of $15,000/part. How many spares should be bought per bus now to support 10 years worth of operation? Hint, solve Problem C.9 first.

7.4 Consider the bus problem modeled in Sections 4.4.1.2 and C.2.3. In this case, a constant failure rate with $\lambda = 1.4 \times 10^{-5}$ failures/mile, and 2740 miles of lost usage per corrective maintenance event. Assume there will be no future opportunity to procure additional spare parts. How many spares should be bought per bus now to support 10 years worth of operation? Assume that the part costs $2000 to procure. Assume if you run out of parts you will have to procure the part from a third party at an effective cost of $20,000/part. Assume that the part is new at the start of the 10 year period. Determine the optimal lifetime buy quantity.

(a) If the cost of money (WACC) is 10%/year.

(b) If the cost of money (WACC) is 10%/year and the holding cost of parts is $1000/year/part.

Hint, start with the simulation described in Section C.2.3 and extend it to compute cost.

7.5 Using the Porter model, what year should a design refresh be performed if $C_{DR_0} = \$67{,}000$, $r = 22\%/\text{year}$, $Q_i = 500$ (for 15 years and zero thereafter), $C_h = 0$, and $P_0 = \$16/\text{part}$?

7.6 Find the optimum refresh year using Equation (7.17) for one part with $Y_O = 0$, $C_{DR_0} = \$100{,}000$, $r = 12\%/\text{year}$, $Q_i = 500$ (for 20 years and zero thereafter), $C_h = \$1/\text{part/year}$, and $P_0 = \$10/\text{part}$?

7.7 Find the optimum refresh year using Equation (7.17) for a system composed of two different parts:

(a) Each having the same data given in Problem 7.6.

(b) Each having the same data given in Problem 7.6, except one of the parts has $Y_O = 3$.

Chapter 8

System Sustainment Enablers

In Section 1.4.4, of this book the concept or resilience was discussed. Resilience is the ability of a *system* to resist disturbances. The preceding chapters of this book have focused on the acquisition, hardware and software reliability, and various attributes of logistics associated with sustaining systems. In this chapter we look briefly at several other elements that are essential to creating and maintaining system resilience including: workforce aging, technology management, system end-of-support, and other topics.

8.1 Workforce Aging Management

Workforce aging impacts both the workers in a workforce and the organizations that they work for. The effect of workforce aging on organizations includes: productivity, stability, retention of knowledge, retention of workers, and in some cases the ability to continue the support of critical systems. The magnitude and impact of workforce aging depends on the particular product and/or service sector that an organization is engaged in. For some sectors, the loss of worker skills and experience can be mitigated via knowledge capture/transfer and the training of younger workers, however, in other sectors, experience can be very difficult to replace, once it is lost.

Obsolescence (described in Section 7.5.1), is not just a hardware and software problem; it also impacts the availability of the human skills necessary to manufacture and maintain systems.

Organizations that manufacturing and support systems require a pool of specialized skills in order to perform their processes without interruption. However, the pool of available workers and their associated skills changes over time. Workforce planning is commonly defined as getting the right number of people with the right competencies in the right jobs at the right time. There is a myriad of issues associated with mismatches between the skills possessed by the workforce and the skills needed by the organizations that manufacture and sustain systems. These issues can be classified into the following three categories: skills obsolescence, skill shortage, and critical skills loss.

- *Skills obsolescence* (also referred to as human capital obsolescence) is the situation where there is an adequate number of workers, but they do not have the appropriate skills and need to be retrained. Skills obsolescence includes situations where the supply of an old skill exceeds the current and expected future demand for that skill. A large amount of literature exists on the analysis of skills obsolescence, which is not particularly relevant to the sustainment of systems.
- *Skill shortage* means that there are insufficient skill competences in the workforce, i.e., the current and future demand for a skill exceeds the supply of that skill. As a result, there is a need to identify, train, and retain the workforce to fill current and expected future skill gaps.
- *Critical skills loss*, is the process that leads to skill shortage.

The modeling of critical skill loss and its impact on application-specific skill shortage, is the topic of the remainder of this section. Critical skills loss, as defined in [8.1], describes the loss of skills that are either non-replenishable or take very long periods of time (possibly years) to reconstitute. Critical skills loss is a special case of "organizational forgetting", which describes the situation where organizations lose knowledge gained through learning-by-doing due to labor turnover, periods of inactivity, and/or failure to institutionalize tacit knowledge [8.1]. Most analyses of organizational forgetting assume that the situation is relatively short term and seek to forecast the recovery period and the associated disruption in productivity and schedule. However, critical skills loss, as used it in this book, represents a long-duration and involuntary form of organizational forgetting. The type of critical skills loss encountered when sustaining systems is usually the result of long-term attrition where skilled workers retire and there are an insufficient number of younger workers to take their place. It is key to note that critical skills loss is not the result of inactivity, poor planning, or a lack of foresight by an organization, but rather simply the inevitable outcome of the organization's dependence on specialized skills for which there is relatively little demand (which is also the nature of DMSMS-type obsolescence problems for hardware and software discussion in Section 7.5.1). Various system support and management challenges created by the loss of human skills have been reported in industries including: healthcare, nuclear power, defense, and aerospace.

Suggested causes for critical skills loss include: declines in education and training (e.g., university educated STEM students are no longer trained in the programming languages, such as COBOL, still used in some legacy systems); younger workers shy away from entering particular workforces that they perceive to be in decline (e.g., nuclear power) or not cutting-edge; younger workers leaving legacy system jobs to pursue what they perceive to be more lucrative and exciting opportunities; shrinkage of "feeder" occupations (e.g., the U.S. Navy feeding skilled workers to the nuclear power industry); adverse demographics (e.g., workplaces in geographical areas that people do not want to live); older workers with critical skills aging out and taking irreplaceable tacit knowledge with them (or in some cases, older workers unwilling to transfer what they know to protect their jobs); and differences in social and cultural influences between younger and older workers regarding the perceptions of their jobs.

As the workforce that possesses the required skills to support a system shrinks, the amount of time that the system is down (non-operational) when it requires support increases leading to reduced availability, which can lead to a loss of revenue, safety compromises, property damage, loss of mission, and loss of life.

8.1.1 *Workforce pool forecasting*

Management of the workforce available to sustain a system requires the ability to predict the following: (1) the workforce pool size as a function of time; and (2) the equivalent skill level in the pool. Sandborn *et al.* [8.1] gives the total number of people in the workforce pool for age a in year i as,

$$n_i(a) = [n_{i-1}(a - 1) + H_i P_0 f_H(a)](1 - f_L(a)) \qquad (8.1)$$

where H is the hiring rate per year relative to the pool size at time 0, P_0. f_H is the hiring age distribution and f_L is the exit age distribution. In the simplest model, the hiring and exit age distribution can be assumed to be constant. A constant f_H does not imply that the same number of people are hired every year, it only implies that the people that are hired have ages distributed like f_H. The same assumption is made for f_L, i.e., the number of people that exit is not the same every year, but their distribution is always f_L. The f_H and f_L distributions (and P_0) are determined from the workforce pool at the current time. The initial distribution n is found from,

$$n_0(a) = P_0 f_C(a) \qquad (8.2)$$

where f_C is the distribution of the current ages of the workers in the pool.

To demonstrate how this workforce model works, consider an organization that has the attributes given in Table 8.1.

Assuming that the current pool size (P_0) is 1000 and workers can remain in the workforce until age 80. The results of the model are shown in Figure 8.1 for two different hiring rate assumptions. The result shows that without any hiring ($H = 0$) the total workforce pool size will drop to zero within 55 years. If 10% of the workforce can be rehired per year, the workforce will grow for approximately 10 years and then begin to drop, and drop below the current workforce size in approximately 30 years.

Calculating the size of the workforce pool (head count) over time is necessary but not sufficient to capture an organization's future ability to support a system because workers have different levels

Table 8.1. Workforce example characteristics (Weibull distributions assumed).

Hiring Age Distribution (f_H)		Exit Age Distribution (f_L)		Current Age Distribution (f_C)	
Scale (η)	10 yr	Scale (η)	24 yr	Scale (η)	23 yr
Shape (β)	2	Shape (β)	4	Shape (β)	2
Location (γ)	22 yr	Location (γ)	22 yr	Location (γ)	22 yr

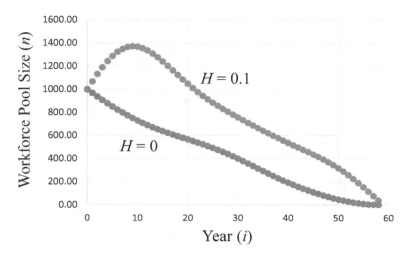

Fig. 8.1. Workforce size as a function of years into the future for two different hiring rates.

of skill. Because of the varying skill levels, not all workers provide an equivalent level of "value" to the support of the system. The cumulative skill must be estimated in order to track the value (productivity) of the pool of workers. In this model, skill level is defined as the length of time that a worker has spent in a particular position.[1] The cumulative skill level in the workforce pool in year i, S_i, is calculated using,

$$S_i = \sum_{a=a_y}^{a_o} (aP_1 + P_2)\frac{n_i(a)}{P_0} \tag{8.3}$$

where, P_1 and P_2 are organization-specific parameters used map age to the skill level measured in years, a_y is the age of the youngest worker in the skills pool at the start of the analysis (from the f_C distribution) or the youngest worker that is hired during the analysis period (from the f_H distribution) and a_o is the age of the oldest worker (e.g., the mandatory worker retirement age). Note, while "skill level" has the units of time, S_i, which is used in the model, represents the cumulative skill relative to the initial condition.

---

[1]Other measures of skill may be applicable depending on the context.

System Sustainment

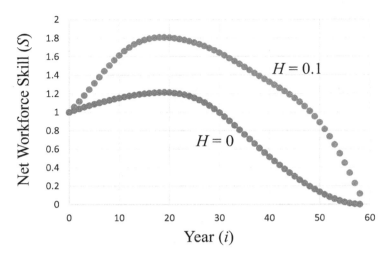

Fig. 8.2.   Workforce skill as a function of years into the future for two different hiring rates.

Using Equation (8.3) the time to perform an activity (e.g., a maintenance activity) in year i could be given by

$$T_i = \frac{T_0}{S_i} \tag{8.4}$$

where T_0 is the time to perform the activity with a skills pool having S_0 skill. In Equation (8.4) the time required to perform the activity increases as skill decreases due to the following factors: 1) less-skilled workers require more time to perform the activity (learning curve effects); and/or 2) if the pool of workers capable of performing the required activity shrinks, appropriate workers may not be available at every site and may have to travel from a different location, which takes time.

Using parametric values of $P_1 = 0.8808$ and $P_2 = -20.905$ (from [8.1], which are application specific), and Equation (8.3), the total pool skill as a function of time for the example case given above is shown in Figure 8.2.

Note, that even with no hiring, the skill level of the workforce initial increased (the existing workforce is gaining skill faster than the people are leaving). In this case, even with no hiring, the net skill level of the workforce does not drop below the current level for 30 years.

However, this does not mean that the skill remains geographically distributed in an appropriate way, i.e., one super-skilled person may not be able to do the job of several lesser skilled workers if the skills need to be spread around different sites.

8.2 System Redesign (and Refresh)

During a system's life cycle, system refresh may be desirable (or necessary) to assure the system remains supportable, to maintain its relative capability so it can continue to be competitive, and/or comply with a system evolution/upgrade roadmap.

Refresh planning (discussed in Section 7.5.1.3) replaces obsolete or technologically out-of-date components with non-obsolete parts. A simple refresh of this type is performed to keep the system manufacturable and/or supportable (its primary goal is not to improve performance or functionality).

In some cases, systems have requirements or roadmaps that dictate that the system (hardware and software) must be upgraded in specific ways at specific points in time (we call this a "redesign"). These types of system upgrades usually support a combination of performance enhancements and functionality enhancements to the system. If/when these types of system updates occur, obsolete parts, and potentially problematic parts,[2] will also be addressed.

8.2.1 *System capability refresh*

Capability is defined as "the ability to achieve a desired effect under specified standards and conditions through combinations of means and ways to perform a set of tasks" [8.2]. For the purposes of this discussion, we will define the system's technological capability as its ability to accomplish the "mission" or purpose it was designed for. For example, the absolute capability of a fighter aircraft radar system is its effectiveness in detecting objects in the surrounding area, while its relative capability is its effectiveness detecting adversaries early

---

[2]Problematic parts are those that have encountered reliability issues in the field, or possibly quality issues during manufacturing.

enough so that appropriate action can be taken. A system for which relative capability does not erode over time (because the adversary does not change over time) might be a weather radar system.

If the system's relative capability is adequately maintained without upgrading the system, then refresh planning (Section 7.5.1.3) becomes a life-cycle cost minimization problem. However, if there is no externally mandated system upgrade strategy, and relative capability erodes without upgrades, then one needs a way to evaluate the impact of the erosion in relative capability in order to optimally plan refreshes.[3] Note, the concept of capability relative to an adversary, while obviously applicable to defense applications, is also applicable to commercial applications where the adversary is the competition. Commercial systems must maintain a capability relative to their competition or they will lose market share. A system for which relative capability does not erode over time (because the adversary does not change over time) might be a weather radar system.

There are several approaches to this problem that can be taken. One is to assume that value lost is a cost ("capability cost") and minimize the combination of refresh costs and capability costs, e.g., a defense system. Alternatively, one can seek to maximize the lifetime value provided by the system, while concurrently minimizing its life-cycle cost, e.g., a weather radar.

Considering capability as a cost, the cost of system technological capability is not just the cost to implement the capability (which certainly needs to be accounted for), but, more importantly, the costs that result from the lack of capability. More precisely, the cost is a result of the effectiveness of the system in performing the tasks required of it. For example, failure to upgrade the capability may result in less mission effectiveness or a higher probability of losing the asset due to the actions of an adversary – these may translate into costs (e.g., fleet availability or fleet effectiveness costs).

Figure 8.3 shows conceptually how the life-cycle capability cost can be evaluated based on the capability gap between an example system and the state-of-practice of the technologies that comprise the system during the system's operational life. The shape of

---

[3]This is a version of dynamic valuation [8.3]. Dynamic value is an estimate of how the value of a system fluctuates its lifetime.

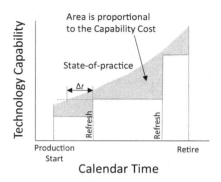

Fig. 8.3. The relationship between the system's technological capability and capability cost [8.4].

the state-of-practice curve in Figure 8.3 depends on the relevant technologies and their evolution over time. State-of-practice represents technology that has matured sufficiently to be practical for incorporation into relevant systems (e.g., it may be the level of technology that an adversary is using). Δt represents the technology lag time, which is the time it takes to implement a design refresh in the system (design, qualification, and fielding). The larger the Δt, the more likely a system will lag in a capability competition. The area between the state-of-practice curve and the system capability curve is proportional to the total cost of capability during the system's operation life.[4]

As shown in Figure 8.3, each delivery of a technology refresh to the fielded system instance resets its capability to a higher level, closing the gap between the system and the state-of-practice capability. Frequent technology refreshes keep the system up-to-date during its life cycle, reducing the probability of losing the technology competition and decreasing the corresponding capability cost. Note, Δt could be negative if a refresh pushes the capability of a system ahead of the adversary. Also, the Δt will be system instance specific if not all

---

[4]Figure 8.3 assumes that an adversary's capability distribution shifts with state-of-practice. It is possible that the adversary is not bound by the same state-of-practice that is relevant for the example system, e.g., terrorists may use technology that has not matured sufficiently to be incorporated into traditional defense systems shifting the curve up from that shown in Figure 8.3.

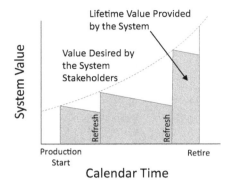

Fig. 8.4.   The relationship between system value and lifetime value provided by the system [8.3].

fielded instances of a system receive a refresh at the same time, e.g., hardware refreshes may take years to reach all the fielded instances of a system.

Figure 8.3 provides a conceptual evaluation of capability cost. Obtaining the quantitative cost, requires constructing a stochastic relationship between capability and cost, see [8.4]. Zellers [8.5] provides a similar capability evaluation where capability is called system value and the state-of-practice is the desired performance of the system, and the actual performance is the requirements.

Figure 8.4 shows an alternative model proposed in [8.3]. In this case the value provided by the system decreases linearly after a refresh is completed. In the model shown in Figure 8.4, one seeks to maximize the shaded area concurrent with minimizing the life-cycle cost of the system.

For both of the models described in this section, the overall objective is to determine when to refresh the system, i.e., how many refreshes should be done and where (in time) those refreshes should be located.

8.3   Configuration Management

While all the fielded instances of a critical system are, in theory, the same system, in reality they are not. When the instances rolled off the manufacturing line, they might have been the same, but as they are maintained, refreshed, etc., they may all migrate to slightly (or not so slightly) different configurations.

Configuration management is a component of project management that focuses on defining how to manage changes to system configuration. A configuration includes the physical and functional specifications of a system.

Configuration control is where the actual changes to the system are managed throughout the system's life cycle. The actual configuration control process is not in the scope of this book, rather, our interest is in the resources required to sustain a system that has many different configurations.

F-35 Variants

The F-35 has three main variants: the conventional takeoff and landing (CTOL) F-35A, the short take-off and vertical-landing (STOVL) F-35B, and the carrier-based Catapult Assisted Take Off Barrier Arrested Recovery (CV/CATOBAR) F-35C. There is also an F-35I that is an F-35A with unique Israeli electronic warfare systems, including sensors and countermeasures. Two additional variants have been proposed, and F-35D, which would be an upgrade to the F-35A to support the USAF's Future Operating Concept; and a CF-35 that has a drogue parachute and may include an F-35B/C-style refueling probe.

Within these variants, there are other variations, for example there are F-35As with drag chute pods and F-35As with probe-and-drogue refueling modifications.

The variants discussed above represent the "factory delivered" versions of the F-35. The fielded F-35s will, in most cases, be upgraded with various technology refreshes of their avionics over their usage life proliferating the factory variants. For example, Marine F-35Bs and the Navy F-35Cs began to receive incremental Block 4 upgrades in 2021. The F-35's Block 4 upgrade provided more computing power, processing power, and memory than the previous F-35 computers. Block 4 software upgrades also improved the flexibility and enabled the F-35 to operate AI-flown wingmen, as well as improve the F-35B and C's electronic warfare capabilities in jamming enemy radar and transmissions.

8.4　System Test and Qualification

A key element of system acquisition is the testing and qualification of the system. As a result, the development team should develop a comprehensive program test plan early in their process, which defines the objectives and scope of the testing effort. The plan identifies the methodology to conduct tests, the hardware, software, and tools required, as well as the functions to be tested. During development, testing can provide data to assist developers in assessing the level of performance, identifying and correcting deficiencies, and support trade-off analysis.

In general, testing during a system's production falls under two separate categories: (1) recurring functional test during system manufacturing and installation;[5] and (2) reliability testing to establish the reliability and perform qualification/requalification of the system. Reliability testing (also called life testing or environmental testing) is performed to provide confidence that the system meets its reliability requirements. Reliability testing is performed before products or systems are put into the market or the field. Qualification verifies that the reliability achieved by the system meets the specified requirements for the system.

8.4.1　*Recurring functional test*

Recurring functional testing is done as part of the manufacturing/ installation process for every instance of the system. The outcome of recurring functional testing is: a pass followed by the system instance's continuation in the manufacturing/installation process, a fail followed disposal of the system, or a fail followed by diagnosis/rework/retesting. The goal of functional test strategies is to answer the following questions:

1. When should a system be tested? At what point(s) in the manufacturing process?

---

[5]Note, all or part of the recurring functional testing performed as part of the manufacturing and/or installation process may be required during the maintenance process to ensure that the repaired system meets functional requirements.

2. How much testing should be done? How thorough should the testing be?

3. What steps should be taken to make the system more testable?

Limited resources (time, facilities, and money) require the determination of how to obtain the best test coverage possible for the least cost. The specific goal of test economics is to minimize the cost of discarding good products and the cost of shipping bad ones. This goal is enabled through the development of models that allow the yield and cost of products that pass through test operations to be predicted as a function of both the properties of the product entering the test and the characteristics of the test operation (its cost, yield, and ability to detect faults in the product it is testing).

The optimization of recurring functional testing is part of the manufacturing process and will not be addressed further in this text.[6]

8.4.2 *Environmental testing*

A full reliability testing program includes various categories of test (functional, environmental, etc.), however, here we will focus on environmental testing, which is performed to establish (or confirm) the reliability of the system. Environmental testing consists of tests that subject the system to appropriate environmental conditions (or combinations of conditions) until the system fails. The word "appropriate" is important because there may be little value in subjecting the system to conditions (stresses) that it will not encounter in its field life or do not accelerate relevant failure mechanisms.

Environmental testing is intended to cover the range of environmental conditions that the system is likely to have to endure during its life. The primary conditions affecting most system (more specifically the conditions that contribute to the accumulation of damage against relevant failure mechanisms for systems) include:

- Temperature (hot and cold extremes)
- Thermal Shock
- Vibration
- Electromagnetic effects (EMI)
- Radiation
- Dust/dirt (contaminants)

---

[6]Acceptance testing is a test conducted to determine if the requirements of a specification or contract are met. Acceptance testing usually occurs at or before the customer takes ownership of the system.

- Mechanical Shock
- Humidity
- Altitude (pressure)
- Vacuum (lack of atmosphere)

- Salt Spray
- Gasses (including pollution)
- Fungus
- People

There are, of course, other more specialized environmental conditions that can contribute to some relevant failure mechanisms in some types of systems. The environmental testing attributes of reliability testing must determine which environmental conditions (taken singularly, or in combination), are likely to be the most critical from a system life point of view. The specific methods of designing and applying the environmental tests are outside the scope of this text, but many reliability engineering texts focus on this for various system domains.

8.4.3 *Accelerated life testing*

Because it is often impractical to run environmental testing in real time, i.e., it takes too long to subject the system to actual conditions for an entire system life to assess reliability, often accelerated testing methods are used. Accelerated life testing subjects a product to stress conditions (temperatures, voltage, vibration, humidity, pressure, etc.) in excess of the conditions it is expected to experience in the field in order to precipitate faults and potential failure modes in a shorter amount of time, i.e., to accelerate failures. By analyzing the product's response to accelerated test conditions, one can make predictions about the service life and maintenance intervals for the system under actual life conditions.

The concept of accelerated testing is that higher stress levels will shorten the expected system life by increasing failure rates. However, a great deal of care must be taken to design accelerated tests that actually accelerate relevant failure mechanisms without introducing new mechanisms that are not relevant to the life of the system.

Accelerated life testing is commonly characterized using an acceleration factor defined by

$$AF = \frac{L_{Field}}{L_{Test}} \tag{8.5}$$

where L_{Field} is the system's life in the field and L_{Test} is the system's life at the accelerated test (stress) level. This relationship assumes

Table 8.2. Relationship between acceleration factor (AF) and reliability functions.[*]

Reliability Function	Relationship
Time to failure, t	$t_{Field} = (AF)t_{Test}$
Failure probability density function, $f(t)$	$f_{Field}(t) = \dfrac{1}{AF}f_{Test}\left(\dfrac{t}{AF}\right)$
Failure cumulative density function, $F(t)$	$F_{Field}(t) = F_{Test}\left(\dfrac{t}{AF}\right)$
Reliability, $R(t)$	$R_{Field}(t) = R_{Test}\left(\dfrac{t}{AF}\right)$
Hazard rate, $h(t)$	$h_{Field}(t) = \dfrac{1}{AF}h_{Test}\left(\dfrac{t}{AF}\right)$

[*]Time (t) is used in this table, but the relevant stressing factor can be substituted for it.

that the same failure mechanism is caused by the same environmental stress under both field and test conditions. Equation (8.5) can be used to determine the test time (or whatever equivalent stressing units are relevant) that is equivalent to a specified field life.

As an example, consider testing a mechanical part that is assumed to have a constant failure rate. Assume that the dominant failure mechanism for this part depends simply on time at temperature. Assume that a doubling of the temperature is known to have an acceleration factor (AF) of 14.6. If an accelerated life test run at 120°C determines that the part has an *MTBF* for 1000 hours, what is the expected *MTBF* of the part under field conditions for 60°C assuming 2 hours of use per day? From (8.5), the test life is, $L_{Field} = (14.6)(1000\,\text{hours}) = 14{,}600\,\text{hours}$. At 2 hours per day the field life is $(14{,}600)/(365\,\text{days/year})/(2\,\text{hours/day}) = 20$ years. So, the *MTBF* of this part in the field is 20 years.

The effect of the acceleration factor (AF) on the various reliability functions introduced in Chapter 3 is given in Table 8.2.

Acceleration models have been developed associated with various different stresses (e.g., temperature, humidity, vibration, etc.). These models are characterized by various empirical constants determined from testing that correspond to different types of parts, materials, failure mechanism, operating regimes, etc.

Depending on the nature of the accelerating stress factors, it is possible to misinterpret the results of accelerated life tests. The implementation of meaningful accelerated reliability tests requires an accurate understanding of the proper accelerating stresses and the design limits of the product.

8.4.4   *Maintainability demonstration*

Maintainability demonstration is conducted to verify that a system's maintenance requirements have been met during fielded operation. Maintainability demonstration involves the selection and performance of maintenance tasks based on the task's expected contribution to the total maintenance time/cost. The techniques are essentially the same as for maintenance prediction, i.e., Section 4.3, but the task times are measured rather than estimated. Depending on the particular maintainability measure you wish to demonstrated, a test plan is conducted, and once a statistically significant amount of data is collected, the data is used to determine if the maintainability is acceptable or not [8.6].

8.4.5   *Acceptance tests and certification*

The results of the qualification tests are used to develop the acceptance tests for the production items, to ensure that these products meet the design requirements and specification, and to ensure the production item has the same performance as the qualified design. These tests are performed by the customer personnel. Data from these tests are collected to begin to develop the reliability and maintainability database. This data can be used to project the system's operational reliability and maintainability, and ultimately the systems availability. The test results are also critical for the identification of design deficiencies and failure modes, project failure rates, project the number of spares required, and to refine the maintenance processes. As systems mature, especially as they approach the end of their useful lives testing can be used to help inform the decision whether to upgrade the system or develop a replacement.

Certification is similar to qualification but is performed by an independent third-party agency to ensure that the product, system, or service meets a specific set of requirements. A written assurance (i.e., a certificate), is then issued. For example, the Federal Aviation

Administration's (FAA) maintains an aircraft certification processes to assure safe aircraft designs. As part of their process the FAA reviews any proposed designs and the methods that will be used to show that the design complies with FAA standards; conducts ground and flight tests to demonstrate that the airplane meets the FAA standards; evaluates the airplane's maintenance and operational suitability for introduction of the aircraft into service; and works with other civil aviation authorities to secure their approval of the aircraft [8.7].

8.5   Cost–Benefit Analysis[7]

Decision making for system sustainment is based on many factors. Making business cases to spend money on strategic management solutions requires approaches that balance costs against the benefits received from the investment. Sometimes the return on the investment can be articulated as financial value and sometimes the benefits are difficult (or impossible) to translate into money. For many public enterprises, it is difficult to justify spending money based solely on return on investment arguments. In these cases, the decision to spend money has to be based on more than just economics. CBA provides a framework to assess the combination of costs and benefits associated with a particular decision or course of action.

Ideally cost–benefit analyses take the broadest possible view of costs and benefits, including indirect and long-term effects, reflecting the interests of all stakeholders affected by the program. If all relevant benefits are simply increases in revenue or cost savings, then CBA is not necessary – a simple cash flow analysis or ROI will suffice.

The idea of CBA is usually attributed to Jules Dupuit, a French engineer in the mid-1800s [8.8]. The practical development of CBA came as a result of the Federal Navigation Act of 1936 [8.9]. This act required that the U.S. Corps of Engineers carry out projects for the improvement of the waterway system in the United States when the total benefits of a project exceed the costs of that project, which made it necessary for the Corps of Engineers to develop systematic methods that enabled the concurrent measurement of benefits and costs.

---

[7] Also known as Benefit–Cost Analysis (BCA) and Benefit/Cost (B/C) Analysis.

CBA is used when the benefits are not monetary, but can be monetized. It is the process of monetizing the non-monetary benefits that makes CBA challenging. An example CBA analysis is provided in Section 8.5.4, however, first we need to introduce several additional concepts.

8.5.1   *Cost avoidance*

In the case of manufacturing processes, it is reasonable to characterize the value of process, equipment, and yield changes as cost *savings* (or a cost reduction). However, the value of sustainment activities is usually quantified as a cost *avoidance*. "Cost avoidance is a cost reduction that results from a spend that is lower than the spend that would have otherwise been required if the cost avoidance exercise had not been undertaken" [8.10]. A simpler definition of cost avoidance is a reduction in costs that have to be paid in the future to sustain a system.[8]

The reason that cost avoidance is used rather than cost savings, is that if the value of an action is characterized as a cost savings, then someone wants the saved money back (or there is an opportunity for a budget cut). In the case of sustainment activities there is no money to give back (and there may not have been any budget in place to begin with). Unfortunately, making business cases based on a future cost avoidance argument is usually more difficult than business cases that are based on cost savings, therefore, there is a greater need to be able to provide detailed quantification of sustainment costs. Requesting resources to create a cost avoidance is not as persuasive as making a cost savings or a return-on-investment argument.

The most general cost avoidance calculation is,

$$Cost\ Avoidance = Life\ Cycle\ Cost_{original} - Life\ Cycle\ Cost_{new} \tag{8.6}$$

where original refers to the projected life-cycle cost for the system without the new process, management or technology incorporated,

---

[8]Other similar definitions exist, e.g., "a cost reduction opportunity that results from an intentional action, negotiation, or intervention" [8.11]. Note, the term "cost avoidance" is also used in the context of health care.

and *new* refers to the life-cycle cost of the system after the new process, management or technology is incorporated. This calculation is used in the next subsection and in Section 4.6 in the numerator of the return-on-investment calculation for preventative maintenance.

Cost avoidance can be quantified in the context of many different activities. Consider one example: design reuse of software. For example, in this case, cost avoidance is given by [8.12]

$$Cost\,Avoidance = C_{cc}(E_n - E_r) - C_{COTS} \qquad (8.7)$$

where C_{cc} is the cost of new custom code (in personnel hours). E_n is the estimated effort without reuse, E_r is the estimated effort with reuse, and C_{COTS} is the direct cost of buying the component. As an example, consider a software project that would require $E_n = 40,000$ hours if developed from scratch, but 5000 development hours can be avoided by buying a commercially available library (COTS), so with reuse, $E_r = 40,000 - 5000 = 35,000$ development hours. Assume the cost of developing new code is \$100/hour and the cost of purchasing the library is $C_{COTS} = \$100,000$. We can calculate the cost avoidance associated with purchasing the library as,

$$Cost\,Avoidance = (100)(40,000 - 35,000) - 100,000 = \$400,000$$

In the next subsection, the calculation of a cost avoidance-based return on investment is discussed.

8.5.2 *Return on investment (ROI)*

Return on investment assumes that the organization has already committed resources to a course of action and seeks to determine if the investment of those resources will pay off.

ROI is a common performance measure used to evaluate the efficiency of an investment or to compare the efficiency of different investments. To calculate *ROI*, the benefit or gain associated with an investment is divided by the cost of the investment and the result is expressed as a percentage or a ratio:

$$ROI = \frac{Return - Investment}{Investment} \qquad (8.8)$$

An *ROI* of 0 represents a break-even situation – that is, the value you get back exactly equals the value you invested. If the *ROI* is > 0, then there is a gain; if the *ROI* is < 0, there is a loss. The *ROI* calculated in Equation (8.8) is relative to the no investment case (e.g., burying the cash in the ground). *ROI* corresponds to an IRR (internal rate of return) calculated over the entirety of the investment, without consideration of the time value of money.

In private sector organizations, investments are often classified as either revenue generating, cost savings, or must do projects. A strong bias exists in private sector firms toward investing in revenue generating projects because of the emphasis by investors on revenue growth of companies. Making the business case for cost savings projects particularly difficult relative to revenue generating projects. In revenue generating projects (e.g., investing in a new product line), the ex-post benefits can be observed in terms of the new revenues. Deriving the ex-post benefits from actual cost avoidance investments is far more difficult than evaluating the ex-post benefits from revenue generating projects, since the ex-post benefits cannot be directly observed but must be estimated. In other words, the potential costs associated with the problems avoided are never actually observed. Furthermore, these cost savings involve implicit costs, as well as explicit costs. The explicit costs are easier to quantify; they relate to the costs associated with things such as correcting a maintenance problem. The implicit costs relate to the costs associated with potential legal liabilities, potential lost sales, and, perhaps most importantly, the operational impacts; these can significantly exceed the explicit costs. The DoD faces similar challenges since, within its constrained budgets, these cost saving investments compete with other investments and the explicit costs can include reduced mission effectiveness (i.e., reduced availability or readiness) and potentially mission failure.

Equation (8.8) is trivial, but *ROI* concepts have not been widely (or consistently) applied to cost-avoidance business case arguments.[9] Cost avoidance (see the previous subsection) is a reduction in costs that have to be paid in the future.

Applying Equation (8.8) to cost avoidance can be done. As an illustration, consider the determination of an *ROI* for the addition

---

[9]It does not help that engineers are notorious for grabbing two numbers, almost any two numbers, dividing them, and calling it an *ROI* (inconsistent and ill-defined *ROI*s appear throughout the engineering literature).

of a system improvement (it could be a technology insertion, the addition of system health management, pro-active DMSMS management, etc.). This change/improvement will result in a future cost avoidance (e.g., a reduction in future maintenance costs or improved system availability). Is the investment worth it? The *ROI* from Equation (8.8) becomes [8.13]:

$$ROI = \frac{C_{wo} - C_w}{I_w} \qquad (8.9)$$

where

C_{wo} is the life-cycle cost of the system when managed without the change/improvement

C_w is the life-cycle cost of the system when managed with the change/improvement

I_w is the investment in the change/improvement.

If C_{wo} is larger than C_w then there is a positive ROI. The nice thing about Equation (8.9) is that the numerator is the difference of life-cycle costs, i.e., one only has to be able to determine the difference between the two cases (the absolute values of C_{wo} and C_w, which are much harder to determine, are not needed). Note that C_w is a life-cycle cost that includes I_w within it. For critical systems, life-cycle costs and investments may be made over long periods of time, therefore, all the values on the right side of Equation (8.9) should be present values discounted using the prevailing WACC for the organization.

Returning to the example at the end of Section 8.5.1, the *ROI* for the investment in the library in this case is given by Equation (8.9) as,

$$ROI = \frac{(100)(40,000) - (100)(35,000)}{100,000} = 5$$

8.5.3   *Should cost*

Should-cost is the result of cost modeling done by a customer to determine what it should cost them to purchase a product, service, or system support [8.14]. Should costing is based on the customer's accounting and their understanding of the product or service after all unnecessary costs are eliminated. When a customer knows a product or service should-cost, then they know what they should pay for it.

For many common products, market competition protects the customer from significant overpayments (assuming the customer shops around). However, for unique products and services (e.g., military programs, contracted services, etc.), where there is no market competition that sets prices at fair levels, the should-cost is critical when a customer wishes to guard against overpayment. As a result, the program office uses a team of engineers, production specialists, and logisticians to perform a comprehensive analysis and review of the contractor's production processes and costs. If you know roughly how much you should be paying, then you can assess whether or not there is sufficient savings or avoidance to justify a sourcing effort.

8.5.4   *A CBA analysis of unmanned drones versus aircraft*

In this simple example, we will assess the cost-effectiveness of using drones to accomplish military objectives versus aircraft. In short, drones are less expensive to procure, operate and support than aircraft. Alternatively, it may take more drones operating for more hours to accomplish the same mission as an aircraft. In addition, drone mishap rates (crashes) are more frequent than aircraft, but drone crashes do not result in human fatalities or as much property damage. Table 8.3 summarizes the input data assumed for this example.[10]

In addition, the value of a statistical life (*VSL*) (see the inset box) for F-35 pilots is assumed to be \$9M and the pilot fatality rate per mishap (R) is 0.27 ($R = 0$ for the drone). Using the data above, we can calculate the total cost per flying hour for each of the aircraft,

$$Mishaps\ per\ flying\ Hour = \frac{M_r}{100,000} \qquad (8.10a)$$

---

[10] The data in this example is "loosely" based on reality, i.e., an in-depth CBA of this tradeoff would require more information, more detail, and increased attention to consistency in the information used.

Table 8.3.   CBA input data.

	MQ-9 Reaper	F-35
(C_P) Acquisition cost per unit	$6.48 million	$90.77 million
(O) Operation and support cost per flying hour	$796	$2536
$(Life)$ Average operational lifetime of a unit	20 years	40 years
(M_r) Mishap rate (per 100,000 flight hours)	7.6	2.36
(L) Monetary loss (less the aircraft) per mishap	$4 million	$10 million
(F) Operator fatality rate/mishap	0	0.27
Maximum payload	3800 lbs	13,000 lbs

Assuming that every mishap results in the loss of the aircraft,

$$Cost\ per\ mishap = L + C_P + (VSL)(R) \qquad (8.10b)$$

$$Mishap\ cost\ per\ flying\ hour$$

$$= (Cost\ per\ mishap)(Mishaps\ per\ flying\ hour) \qquad (8.10c)$$

$$Total\ cost\ per\ flying\ hour$$

$$= Mishap\ cost\ per\ flying\ hour + O + \frac{C_P}{Life} \qquad (8.10d)$$

Value of a Statistical Life (*VSL*)

Although there is a deep aversion amongst many people to the idea of placing a monetary value on human life, some rational basis is needed to compare projects when human life is a factor.

A commonly used monetary value of life is called the value of a statistical life (*VSL*), which is based on the following premise: "the *VSL* should roughly correspond to the value that people place on their lives in their private decisions" [8.15]. If asked, most people would say that they will spare no expense to avoid death, however, economists know that the public's actual behavior (job choice, spending patterns, lifestyle choices) don't agree with this statement. For example, there are many occupations in which people knowingly accept increased risks

(Continued)

(Continued)

in return for higher pay – transmission line workers, oil field workers, miners, construction workers, etc. Using the choices that people make, the value that people place on increased risk can be determined.

The *VSL* is the value that an individual person places on a marginal change in their likelihood of death. Note, the *VSL* is NOT the value of an actual life. It is the value placed on changes in the likelihood of death.

While the whole idea of *VSL* may be ethically troubling, simply ignoring the value of life (and economic cost of death) and leaving it out of CBA could result in a substantial underestimation of the value of the benefits associated with many types of projects. A good overview of *VSL* is provided in [8.16].

Using Equation (8.10d), the cost per flying hour for the drone is $4396 and for the F-35 it is $24,912.

In order to actually evaluate a CBA for this case, we need to value the benefits (not just the operating costs). One way to perform the evaluation is to require that both solutions attain the same level of mission effectiveness.[11] If we assuming that is takes four drones operating 920 hours/year to accomplish the same mission as one F-35 operating 300 hours per year, we can calculate the total annual cost for each case.

$$Drone\,cost\,per\,year = (4)(920)(\$4396) = \$16.18M$$

$$F\text{-}35\,Cost\,per\,year = (1)(300)(\$24{,}912) = \$7.47M$$

Obviously, the analysis is highly dependent on the number of drones needed to perform the same mission(s) as one F-35. Additionally,

---

[11]CBA requires the monetization of all effects, alternatively Cost-Effectiveness Analysis (CEA) does not require the monetization of either the benefits or the costs. CEA does not show whether the benefits outweigh the costs, but rather, shows which alternative has the lowest costs (with the same level of benefits). CEA is often applied when the norm for a certain level of safety or availability has been set.

the relatively inexpensive cost of drones (relative to aircraft) is often weighed against their ambiguous place in humanitarian law.[12]

Multi-Criteria Analysis (MCE) is a tool that allows comparing alternative measures on multiple criteria. In contrast to CBA, MCE allows the treatment of more than one criterion and does not require the monetization of all the impacts. MCE results in a ranking of alternatives.

8.6 Analytic Hierarchy Process

An alternative to CBA for multi-criteria decision modeling is Analytic Hierarchy Process (AHP). AHP is a mathematically rigorous, multi-criteria process for prioritization and complex decision-making, originally developed by Saaty [8.17]. AHP offers a framework to define comparative evaluation that can be used to measure alternatives by breaking a problem down into subproblems, then aggregating the solutions of all the subproblems into a conclusion. AHP can help managers and developers combine both objective and qualitative (subjective) information and make informed decisions, whether prioritizing customer needs, sorting product features, making budget decisions involving a variety of tangible and intangible strategic goals, managing conflicting stakeholders, or selecting from alternative initiatives to be pursued. Converting empirical and subjective experiential data into mathematical models is the main distinctive contribution of the AHP when compared to other decision-making approaches. Although the calculations used in AHP are straightforward, they can become more involved when analyzing complex cases, that may have multi-level hierarchies.

When using AHP the first step is to decompose a problem into a hierarchy of criteria, so these can be more readily independently analyzed and compared. As an example of AHP, consider the drone versus aircraft case described in Section 8.5.4. In this case the goal is to identify the best value platform that can perform the mission in the most cost-effective manner. Additionally, criteria (factors) are the value indicators, for this example we identify three: risk exposure,

---

[12]A debate over the ethical use of drones is beyond the scope of this text.

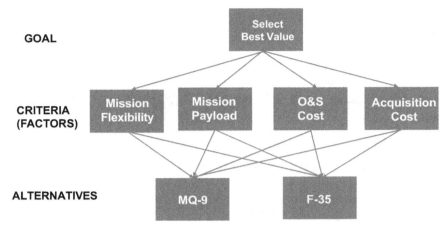

Fig. 8.5.   Hierarchy for AHP analysis of the MQ-9 UAV versus an F-35.

mission performance, and cost. Finally, there are the alternatives we want to consider, in this case there are two alternatives, the MQ-9 UAV and the F-35 manned fighter aircraft. The first step is to build a hierarchy for the decision (Figure 8.5). For this example, the criteria are limited to four items, both objective (Payload, O&S Costs, and Acquisition Costs) and subjective (Mission Flexibility).[13]

The next step is to have decision makers examine the alternatives and make pair-wise comparisons to determine the importance of the criteria with respect to the goal for each of the chosen criteria. These comparisons may use objective data from the alternatives, or qualitative data or judgments as a way to input subjective information [8.18]. The question to ask when comparing two factors is: "Which factor is more important? How much more important is it?" These assessments are generally done using the 9-point scale (Table 8.4).

For example, if we believe that "mission flexibility" is moderately more important than "payload" and "O&S cost", we can use this scale to assign a numeric value of 3 to those comparisons. In a similar way, if it is believed that "mission flexibility" is strongly more important than "acquisition cost" and we assign a numeric value of 5 to the comparison. The principal diagonal of the matrix compares

---

[13]The analysis can be much more comprehensive with more factors and additional hierarchy levels if factors can be further decomposed.

Table 8.4. Saaty's scale of relative importance [8.17].

Comparison Evaluation	Numerical Value
Extremely Important	9
	8
Very Strongly More Important	7
	6
Strongly More Important	5
	4
Moderately Important	3
	2
Equally Important	1

the criteria to itself and is therefore is always of equal importance (i.e., 1). Corresponding entries are a reciprocal, i.e., when "mission flexibility" is evaluated as 3 compared to "payload", then when comparing "payload" to "mission flexibility" the value of 1/3 is used (see Table 8.5). These comparisons are transformed into numeric values that can then be further processed[14] and compared. The criteria weights can be interpreted as the relative importance of each factor with respect to the overall importance of all criteria. In this example "mission flexibility" clearly has the greatest impact, and the "acquisition cost" has the least impact.

For each subjective factor a pairwise comparison is done between the alternatives, and a priority vector is calculated in the same fashion as in Table 8.5. In this example the mission flexibility is the only subjective factor, and this criteria in the F-35 is rated as strongly more important as shown in Table 8.6.

---

[14] A geometric mean is calculated across the horizontal row then normalized to calculate the priority vector. For each row in Table 8.5 the priority vector for a row is given by

$$Priority\ Vector = \left(\prod_{i=1}^{n} x_i\right)^{1/n}$$

where n is the number of columns. The priority vectors for each row are then normalized (by dividing them by the sum of all the unnormalized priority vectors) so that they sum to 1. The result is the Eigenvector of the pairwise comparison matrix.

Table 8.5.　Criteria pairwise comparison matrix.

	Mission Flexibility	Payload	O&S Cost	Acquisition Cost	Priority Vector
Mission flexibility	1.00	3.00	3.00	5.00	0.5174
Payload	0.33	1.00	2.00	3.00	0.2375
O&S cost	0.33	0.50	1.00	3.00	0.1680
Acquisition cost	0.20	0.33	0.33	1.00	0.0771

Table 8.6.　Pairwise comparison of "Mission Flexibility" matrix.

	F-35	MQ-9	Priority Vector
F-35	1.00	5.00	0.8703
MQ-9	0.11	1.00	0.1297

Table 8.7.　Criteria comparison matrix (some data from Table 8.3).

	Mission Flexibility	Payload (lbs)	O&S Cost ($/hr)	Acquisition Cost (M$)
F-35	0.87	13,000	2536	90.77
MQ-9	0.13	3800	796	6.48

For the other attributes the objective values are used (see Table 8.7). Then the attributes that are desirable are identified, i.e., we want greater mission flexibility and payload, and lower acquisition and O&S costs.

For the final analysis, we normalize the objective values to compare them (see Table 8.8). The normalized values are then multiplied by the criteria weight and summed, which for the F-35 equals $(0.5174)(0.8703) + (0.2375)(0.7738) + (0.1680)(0.2327) + (0.0771)(0.0666) = 0.6783$. The same is done for the MQ-9. As a result of this analysis, we see the F-35 has a higher overall priority.

Using the AHP process with this simplified example, we have done pairwise comparison, using both objective and qualitative criteria and synthesized to first calculate criteria importance and then the overall priorities (preferences) to select the best alternative. In this example the final results were heavily influence by the subjective factors. Since this process is intuitive and fairly straightforward, it can

Table 8.8.   Calculation of overall priorities.

	Mission Flexibility	Payload	O&S Cost	Acquisition Cost	Overall Vector
Criteria weight	0.5174	0.2375	0.1680	0.0771	
F-35	0.8703	0.7738	0.2327	0.0666	0.6783
MQ-9	0.1297	0.2262	0.7673	0.9334	0.3217

be easily adjusted for different scenarios, values, and judgments, and can be used as a precursor to a more detailed CBA.

While there is general consensus that AHP is technically valid and useful, the method has been criticized. For example, there is a possibility of rank reversal. Rank reversal occurs when the addition or deletion of a decision alternative results in a change of the relative rankings of the remaining alternatives. If there are four alternatives and as result of AHP they are ranked in order A > B > C > D, if alternative C is dropped you would expect the order A > B > D, however, with some cases the order may change [8.19, 8.20]. This and other criticisms have been identified, and responded to by advocates of AHP [8.21].

8.7   Life Extension

To perform accurate life-cycle cost estimations and to make optimal decisions regarding the sustainment of a system, the system must have a well-defined service life. For example, if a component is discontinued and the mitigation decision is to lifetime buy the component (see Section 7.5.1.2), the total demand for the part for the rest of the system's life is required in order to determine the quantity in the lifetime buy. Obviously in this case, one needs to not only know the annual demand for the component, but also how many years the system needs to be supported, i.e., the planned service life. Unfortunately, with the prohibitively high costs (and long timeframes), associated with critical-systems development and acquisition, coupled with evolving mission requirements, system lives are routinely extended. Obviously, life extensions to assets greatly complicate system sustainment planning. A review of life-extension issues appears in [8.22].

Life Extension: A-10 Thunderbolt II [8.23]

The A-10 "Warthog" aircraft entered service in 1976 to provide close air support for Army ground operations. The A-10 had a designed service life of 6,000 flying hours. This service life was based upon such factors such as historical data of aircraft flying similar flight profiles, engineering fatigue data from destructive analysis of early A-10 prototypes, and commercial wear standards for the wing, fuselage, and empennage structural components. The A-10s in the fleet approached 6000 flying hours in 1997. Rather than retire the A-10, the USAF chose to continue flying the A-10 believing that a life-extension was more cost effective than a replacement (the actual cost effectiveness of this decision is discussed at length in [8.23]).

A program was started in 2005 to upgrade the remaining A-10A aircraft to the A-10C that has modernized avionics necessary for the integration and use of precision weapons. In 2013, Congress and the Air Force examined various alternative proposals for filling the A-10's role, including the F-35 and the MQ-9 Reaper unmanned aerial vehicle. In 2016, the Air Force delayed the final retirement of the aircraft to 2022 to allow the A-10 to be replaced by F-35s on a squadron-by-squadron basis. In in preparation for re-winging the fleet, the Air Force Material Command brought the depot maintenance line back to full capacity in October 2016, and in June 2017, it was announced that the aircraft "... will now be kept in the air force's inventory indefinitely" [8.24].

There are two facets to analyzing life extensions: (1) should a life extension to an asset be done (i.e., can a business be made for this?); and (2) if an asset has been life extended, how does that impact the ability to sustain it?

8.7.1 *Making life extension decisions*

Decisions to extend the life of a system beyond its original designed service life are a tradeoff between the cost of developing and deploying a replacement system, the potentially escalating cost of sustaining the old system, and the ability of the system to continue to meet the current (and evolving) readiness/availability requirements that are placed upon it.

The costs of extending the life of a system must account for sustainment costs (maintenance and spares) the resolution of aging supply chain problems (component obsolescence management and counterfeit part risk), and the loss of critical human skills. Demand for the system (and spares to maintain the system) must be projected through the life extension. The cost tradeoff is presumably made against a replacement system; however, it could be against simply deleting the "mission" or "role" that the system played.

In some cases, life extensions are made to existing systems concurrent with system changes that are targeted at increasing the system's capability or applicability (breadth of mission). For example, wind farms that reach the end of their design life (usually 20 years) can be "repowered", which means upgrading turbines in the farm to newer, possibly larger, and/or more efficient turbines.

The decision to spend money extending the life of a system is also influenced by asset specificity. In economics, asset specificity is the degree to which a thing of value, which could be a physical item, intellectual property, or a person/workforce, can be readily adapted for other purposes.[15] In the case of critical systems, the assets that are being life extended are much more than just the physical item, they include the facilities and equipment used for sustainment and the workforce used for sustainment.

Jones and Zsidisin [8.22] build a model based on transaction cost economics (TCE) to analyze the initial acquisition and long-term support for a piece of capital equipment, and combine it with the projected mission capability rates in order to assess the value of a life extension to the asset. Mission capability rate is an operational availability that includes only the readiness of assets that are currently possessed at the unit level (e.g., aircraft awaiting maintenance at a depot are not included in mission capability calculations).

---

[15] An asset with high specificity is useful only for a very narrowly defined set of tasks and/or circumstances (locations, environmental conditions, or training). An asset with low specificity has more flexibility to be used for other purposes, and therefore potentially has more value.

8.7.2 *Managing systems that have been life extended (the impacts of life extensions)*

Possible life extensions to systems can be modeled as an uncertain end-of-support date for the system. This can be done by representing the end-of-support date as a probability distribution in a stochastic discrete-event simulator (see Appendix C).

8.8 System End-of-life

System retirements are an afterthought for most critical system life-cycle planners. Two questions arise when we talk about system end-of-life: (1) when should a system be retired? and (2) what actions are required to retire the system once a decision has been made.

The need to retire a system can be driven by many different issues:

- *Cost*: systems may simply become too costly to continue to support or operate. There may be tax implications as well.
- *Safety*: a point comes at which the system is deemed unsafe, i.e., it can no longer be "recertified" for safe operation after repair.
- *Regulation/Policy*: the system may no longer be allowed to operate due to changes in regulations, laws, policies, or politics.
- *Repair Limits*: there may be a maximum limit (set by policy) on how many times a system is repaired before it is retired.

DC-10 SN 47844 [8.25]

The DC-10 was introduced by McDonnell Douglas in 1970. A total of 386 aircraft were produced and delivered. One particular DC-10, with the serial number 47844, was delivered to the Brazilian airline Varig, in 1980. In 1994 Varig sold this aircraft to the Uruguayan airline Pluna, who eventually sold it to Northwest Airlines. Northwest retired their DC-10 fleet in 2007 at which time the aircraft was sold to ATA. ATA used it for charters for a year before going bankrupt. This aircraft was then sold to World Airways, which used it for a couple more years before putting it in storage.

- *Part-Out*: To create a source of spares parts to support other systems in the fleet ("fleet planning") – these are often called "hanger queens", i.e., a grounded aircraft that is kept so that its parts can be used in other aircraft. In some cases, system's parts may be worth more than the whole system ("part-out" the system).
- *Replacement*: the system is replaced by some other system and is therefore no longer needed.
- Exhaustion of the ability to support the system due to lack of parts, facilities, or appropriate workforce.
- *Lack of need* (change of mission): the mission that the asset was originally created for has changed and as a result the system (or fewer instances of the system) are needed.

There are some systems that may have effectively infinite service lives (with appropriate maintenance, upgrade, etc.). These include many major civil and infrastructure systems, e.g., interstate highways (bridges notwithstanding), rail lines, the Lincoln Memorial in Washington DC, etc.

8.8.1 End of maintenance (EOM)

The FAA defines *End of Maintenance* (EOM) as "the moment a site requisition cannot be replenished. This stage change begins with the depletion of limited depot and site spares quantities, followed by service degradation (i.e., loss of redundancy) and ultimately loss of system operations" [8.26]. The last portion of the FAA's definition (the loss of system operations) is what we define as the EOM date for the system. In short, the EOM date is defined as "the earliest date that all available inventories fail to support the demand for one or more specific parts resulting in the loss of system operation". An EOM event is defined for a population of systems but is caused by a specific part on a specific system instance at a specific location.

Additionally, the FAA [8.26] defines *End of Repair* (EOR) as "when hardware product support is no longer available by any means or is cost-prohibitive". We consider the EOR date as "the date that the last repair or manufacturing action associated with a part can be successfully performed". EOR events are part specific if all system instances can draw from all inventories (no inventory segregation) and EOR events may be part and system specific if specific systems can only draw parts from specific inventories.

EOM and EOR can be modeled using discrete-event simulations developed for assessing and optimize lifetime buys (Section 7.5.1.2). In the case of optimizing a lifetime buy, an end-of-support date is fixed and the buy size can be adjusted in order to minimize life-cycle costs.[16] In the case of EOM/EOR modeling, the number of parts at the start of the simulation is fixed (it corresponds to the inventories that you have) and the simulation is run (through time) until the demand for a part cannot be satisfied [8.27]. This problem is essentially a large bookkeeping problem, but with complexity that includes:

- Segregated inventories (not all inventories are available to all system instances)
- Loss in inventory (and non-infinite shelf lives for some components)
- Concurrent analysis for an entire bill of materials
- Stochastic demand
- Design refreshes that modify demand for some parts in some systems
- Salvaged parts from discarded assemblies.

8.8.2   *Decommissioning*

Once an owner decides to decommission an asset, it will enter the disassembly process. The purpose of disassembly is to remove the valuable components and security-sensitive technology from the asset. The removed components, depending on their technical condition, may be placed in spare parts inventories, or returned to the market directly (as surplus equipment); or they may need to be inspected and repaired or overhauled by an approved repair shop before returning to service. Security-sensitive technology (if it is not placed in a spares inventory) may need to be appropriately destroyed of otherwise disposed of.

---

[16] Appropriate penalties for having too many or too few parts, holding costs, and shelf life must be included in the model.

In the case of an aircraft, once the aircraft has permanently lost its airworthiness, it will no longer be considered as an aircraft and may be considered as waste instead. Usually this occurs once the last aircraft owner has sold the aircraft to a dismantling company and all parts intended for re-use have been disassembled. Thereafter, it becomes business waste.

Recyclable wastes will be processed, and batches will be prepared for recycling, and the non-recyclable wastes will be prepared for disposal. Hazardous waste materials must be disposed of appropriately. assemblies.

8.8.3 Fleet retirement

Retiring an entire fleet of systems differs from retiring an individual system. Generally, the motivations for retirement are the same as those discussed above. However, the process of retiring a fleet may be more complex.

A fleet is a set of individual systems that collectively perform a task. For example, the USAF manages most of its fleets using equivalent flight hours (EFH). This measure combines flight hours with usage severity information. For example, a strenuous 1-hour mission may have 1.3 EFH, while an easy 1-hour mission could be 0.8 EFH. As the aircraft are flown, each aircraft's remaining EFH decreases. The general shapes of the equivalent remaining flight hours of the entire fleet are shown in Figure 8.6. If no intervention occurs, an aircraft fleet would see aircraft reaching zero remaining EFH at an approximately constant rate that would look like a ramp (left most profile in Figure 8.6). However, in practice it is impractical to retire

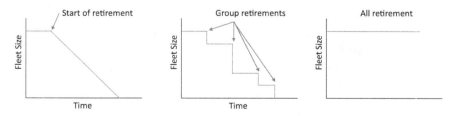

Fig. 8.6.   Fleet retirement strategies. Left: ramp; Middle: multi-step; Right: cliff [8.28].

one aircraft at a time, so like-aged groups are selected for retirement; this retirement pattern is called "multi-step" (middle profile in Figure 8.6). The "cliff" is a profile where all aircraft retire at one time (right most profile in Figure 8.6).

The cliff retirement strategy is the most desirable because it is the easiest to plan for, however, it is also the hardest strategy to achieve in practice – why? Every asset in the fleet has a different number of remaining hours of use in its life. As the assets in the fleet are tasked, they use up the remaining hours that they have. Attaining the cliff strategy in practice requires that increased usage is assigned to assets with less accumulated usage, so that all the assets in the fleet use up all their life at the same point in time.

Optimization of the timeline for the retirement of a fleet can be performed to approximate a cliff. One approach is to task the fleet in such a way as to minimize the total remaining EFH in the fleet at the fleet's retirement time. Newcamp *et al.* [8.28] minimize the following relation for a fleet of aircraft,

$$\sum_{i=1}^{A} \frac{\sum_{m=1}^{M}[\sum_{j=1}^{B}(x_{mj}^i SF_{mj})] + \sum_{j=1}^{B}[(FHR_j^i + AC_j^i)L_{ji}]}{\overline{CSL}_i - IEFH_i} \quad (8.11)$$

where

A	is the number of aircraft in the fleet,
B	is the number of bases where the fleet is assigned,
M	is the number of mission types,
x_{mj}^i	is the number of flight hours flown of type m at base j by aircraft I,
SF_{mj}	is the severity factor for mission m at base j,
FHR_j^i	is the flight hours required for relocation of aircraft i to base j,
AC_j^i	is the administrative cost for relocation of aircraft i to base j (measured in *EFH*),
L_{ji}	is the 1 if aircraft i is assigned to base j, otherwise 0,
$\overline{CSL}_i$	is the maximum service life of aircraft i in *EFH*, and
$IEFH_i$	is the initial *EFH* of aircraft i.

Equation (8.11) allows for each of the A aircraft in the fleet to have a different initial *EFH*, and for those aircraft to be located (and relocated) between B bases. Equation (8.11) also accounts for

M different mission types that the aircraft could be used on. The objective function in Equation (8.11) is minimized by solving for x^i_{mj}. The minimization of Equation (8.11) has to be performed under a number of constraints including: you can't assign an aircraft more *EFH* than it has left, aircraft can only fly missions at the base it is assigned to, negative *EFH* missions are not possible, and there is a maximum limit on the number of aircraft assigned to a base.

References

[8.1] Sandborn, P. and Prabhakar, V. J. (2015). Forecasting and impact of the loss of the critical human skills necessary for supporting legacy systems, *IEEE Transactions on Engineering Management*, 62(3), pp. 361–371.

[8.2] DoDAF. (no date). DM2 Data Groups, https://dodcio.defense.gov/Library/DoD-Architecture-Framework/dodaf20_capability_mm/#:~:text=A%20capability%2C%20as%20defined%20here,published%20by%20the%20Joint%20Staff

[8.3] Engel, A. and Browning, T. R. (2008). Designing systems for adaptability by means of architecture options, *Systems Engineering*, 11(2), pp. 125–146.

[8.4] Chen, S.-P., Sandborn, P., and Lucyshyn, W. (2022). Analysis of the life-cycle cost and capability tradeoffs associated with the procurement and sustainment of open systems, *International Journal of Product Life Cycle Management*, 14(1), pp. 40–69.

[8.5] Zellers, E. M. (2016). *Design of Flexible Technology Refresh Plans for Military Open Systems Architectures*, Ph.D. Dissertation in the Department of Aerospace Engineering, Georgia Institute of Technology.

[8.6] Mil-HDBK-470A Designing and Developing Maintainable Products and Systems, Appendix B Section 3.1.

[8.7] FAA (2020). Airworthiness Certification Overview. Available at https://www.faa.gov/aircraft/air_cert/airworthiness_certification/aw_overview/. Accessed on September 3, 2020.

[8.8] Ekelund, R. B. and Hébert, R. F. (1999). *Secret Origins of Modern Microeconomics: Dupuit and the Engineers* (University of Chicago Press, USA).

[8.9] Fuguitt, D. and Wilcox, S. J. (1999). *Cost-Benefit Analysis for Public Sector Decision Makers* (Quorum Books, USA).

[8.10] Ashenbaum, B. (2006). Defining Cost Reduction and Cost Avoidance, CAPS Research.

[8.11] NASPO, *Benchmarking Cost Savings & Cost Avoidance*, NASPO Benchmarking Workgroup, Sept 2007.

[8.12] Poulin, J. S. (1997). *Measuring Software Reuse: Principles, Practices, and Economic Models* (Addison-Wesley, USA).

[8.13] Feldman, K., Jazouli, T., and Sandborn, P. (2009). A methodology for determining the return on investment associated with prognostics and health management, *IEEE Transactions on Reliability*, 58(2), pp. 305–316.

[8.14] Carter, A. B. and Mueller, J. (2011). Should cost management: Why? How?, *Defense AT&L: Better Buying Power*, Sept–Oct, pp. 14–18.

[8.15] Brannon, I. (2004–2005). What is a life worth? *Regulation*, Winter, pp. 60–63.

[8.16] Viscusi, W. K. (2018). *Pricing Lives: Guideposts for a Safer Society* (Princeton University Press, USA).

[8.17] Saaty, T. L. (1977). A scaling method for priorities in hierarchical structures, *Journal of Mathematical Psychology*, 15(3), pp. 234–281.

[8.18] Saaty, T. L. (2005). *Theory and Applications of the Analytic Network Process: Decision Making with Benefits, Opportunities, Costs, and Risks* (RWS Publications, Pittsburgh).

[8.19] Belton, V. and Gear, A. E. (1983). On a shortcoming of Saaty's method of analytic hierarchies, *Omega*, 11(3), pp. 228–230.

[8.20] Dyer, J. S. (1990). Remarks on the analytic hierarchy process, *Management Science*, 36(3), pp. 249–258.

[8.21] Saaty, T. L., Vargas, L. G., and Whitaker, R. (2009). Addressing with brevity criticisms of the analytic hierarchy process, *International Journal of the Analytic Hierarchy Process*, 1(2), pp. 121–134.

[8.22] Jones, S. R. and Zsidisin, G. A. (2008). Performance implications of product life cycle extension: The case of the A-10 aircraft, *Journal of Business Logistics*, 29(2), pp. 189–214.

[8.23] Keller, J. (2017). Fighter pilot turned Congresswoman throws wrench in quiet plans to cut A-10 squadrons, *Task & Purpose*. June 8. https://taskandpurpose.com/gear-tech/a-10-warthog-fleet-mcsally/.

[8.24] Sellers, A. (2004). *Life Cycle Extension Strategies for Legacy Systems*, M.S. Thesis, Vanderbilt University.

[8.25] Birchall, D. (nd). When does an airliner get retired from use? https://www.quora.com/When-does-an-airliner-get-retired-from-use.

[8.26] Shaffer, G. and McPherson, G. (2010). *FAA COTS Risk Mitigation Guide: Practical Methods for Effective COTS Acquisition and Lifecycle Support Revision 3.2*, Federal Aviation Administration.

[8.27] Konoza, A. and Sandborn, P. (2014). Evaluating the end of maintenance dates for electronic assemblies composed obsolete parts, *ASME Journal of Mechanical Design*, 136(3).

[8.28] Newcamp, J., Verhagen, W. J. C., Santos, B. F., and Curran, B. (2019). Retirement optimization through aircraft transfers and employment, *Journal of Air Transport Management*, 79.

Problems

8.1 Reproduce Figure 8.1 (the workforce size as a function of year into the future). Hint, the implementation of Equation (8.1) is tricky because the $n_{i-1}(a-1)$ is the number of people in the pool in the previous year and for the previous age group.

8.2 In Problem 8.1, what does the hiring rate per year have to be to ensure that you have a minimum of 800 workers in the pool 40 years from the start of the analysis?

8.3 Assume I take a population of parts and test them at 60°C and determine that the *MTBF* of the parts is 500 hours. What is the acceleration factor of the test if I take a similar population of parts and test them at 120°C and find:

(a) The *MTBF* drops to 250 hours

(b) The *MTBF* does not change, what does this result imply?

8.4 Consider testing an electronic part that is assumed to have a single dominant failure mechanism described by a 2-parameter Weibull distribution with $\beta = 1.3$ ($\gamma = 0$). Assume that the dominant failure mechanism for this part depends simply on time at temperature. Assume that a doubling of the temperature is known to have an acceleration factor (AF) of 21. If an accelerated life test run at 120°C determines that the part's scale parameter $\eta = 1000$ hours, what is the expected reliability of this part at $t = 20{,}000$ hours under field conditions for 60°C assuming 2 hours of use per day?

8.5 In the PHM *ROI* example presented in Section 4.6, what if an alternative PHM technology can be implemented that converts all the failures per year to scheduled maintenance events. However, the alternative PHM technology, has a one-time cost of $50,000 per system plus $3000/year/system. Based only on *ROI*, is the alternative PHM technology better or worse than the PHM technology considered in Section 4.6?

8.6 An organization has been supporting a system for several years. The system is repairable and spares are only used to maintain the system while repairs to the original part are made. The repair time for a part is 1.2 months and 512 identical systems are supported. Experience has shown that 9 spares results in a protection level of 0.9015. If the cost of one spare part is $2000 and downtime is valued at $80,000/month, what is the *ROI* associated with buying 9 spares? Ignore the discount rate.

8.7 If the *ROI* over a 10-year period is determined to be 7.1, what is the annual rate (i.e., the IRR) that is consistent with this *ROI*?

8.8 Assuming a capability model like the one shown in Figure 8.3, determine the total capability cost for the following system. Assume that capability is represented by a system performance parameter with the following functional form,

$$\text{State-of-Practice Capability} = e^{at} + b$$

where $a = 0.4\,\text{yr}^{-1}$ and $b = 5$. Assume that the original system is fielded at year 2 and there are technology refreshes at years 4 and 9, then the system is retired at year 15. Also assume that the refreshes all lag the state-of-the-art curve by 1 year ($\Delta t = 1\,\text{year}$). Assume that the discount rate is 0.

8.9 Suppose that the manufacturer of the MQ-9 UAV is able to cut the acquisition cost of the drone in half. Assume that all other parameters describing the drone are the same.

(a) Using CBA, what is the drone cost per year?

(b) Using AHP what are the numerical priorities for the two alternatives?

8.10 Why does AHP use a geometric mean (see footnote 14) instead of any arithmetic mean?

Chapter 9

Contracting for Sustainment

In this chapter, we will be discussing contracts and how they are used when contracting for sustainment.

A contract is a mutually binding legal relationship obligating the seller to furnish the supplies or services and the buyer to pay for them. In the case of sustainment, the sellers are the sustainment providers; we will also refer to them as contractors or suppliers. The buyers are the system operators, and we will also refer to them as operators, owners, customers, or users.

Operators of long-lived critical systems must devote significant resources to operate and maintain their systems. Operators have a range of options when considering how best to sustain their systems and associated components. Although it is possible for operators to use in-house or "organic" resources for these functions, contracting for the sustainment support with an external provider has many potential benefits.

Firstly, when using an external firm, the operator can use competition to identify the providers that will perform the work most efficiently. Additionally, the firms providing the product support will often have employees with the required specific skills, as well as the specialized equipment required. Consequently, using an external contractor is the option often chosen, if not for all, at least for some of a system's sustainment. In some cases, the original equipment manufacturer (OEM) can perform the necessary product support, thereby offering economies of scale. That is, the OEM can use existing production facilities as a source of parts and their experienced personnel for maintenance and repair; i.e., they could then manage the entire supply chain.

9.1 Transactional Contracting for Sustainment

The traditional model for sustaining critical systems is to use a transactional methodology. With this approach, the system operator purchases repair parts or maintenance services from a provider when a repair is needed. In addition, the customer often also specifies in detail the processes to be used and how the result must be delivered. With this approach, there is no incentive for the service provider to reduce the need for repairs and repair parts, and certainly no motivation to develop and implement innovations that may reduce the demand for their services. When the system fails, the provider charges for the repair, or the replacement, on a transaction-by-transaction basis. With transactional sustainment, the more the system fails, the greater the service provider's revenue and workload – value for the service provider is created with each of the transactions.

Current trends, however, have shifted the concept of value from the simple transfer of a specific good or service to the value-in-use or outcome, which is evaluated by the customer. This customer-focused orientation centers on the outcomes created by the goods and activities that are provided [9.1]. Sustainment contracts can then be structured to contract for the delivery of the operator's required outcomes.

9.2 Outcome-based Contracting for Sustainment

The long-term contract structure under which critical systems are delivered and supported plays an increasingly critical role in defining the strategies that govern how sustainment is performed. We will use the term outcome-based contracts (OBCs) to identify this structure. However, depending on the type of system supported this approach is also referred to by a variety of other terms. These include: "performance contracting", "availability contracting", "contract for availability" (CfA), "performance-based service acquisition" (PBSA), "performance-based logistics" (PBL), and "performance-based contracting".

OBCs specify a desired performance outcome, without stipulating that the work be performed in any specific way; the customer only pays for the outcome delivered. As a result, this strategy permits greater flexibility – service providers can innovate to increase efficiency to and reduce the cost of meeting the contract's requirements.

The goal of OBCs is to align the objectives of the service providers with the system's operators so that what benefits the operators can also benefit the service provider, i.e., creating a win-win scenario for both the service provider and the customer. Consider the case of an auto mechanic. Instead of paying for labor on a transactional basis (i.e., only when the mechanic performs a repair in response to a breakdown), the driver contracts with the mechanic for a fixed sum annually to maintain the car in an operational condition, with a specified availability. For example, the agreement could require the vehicle to be available 98% of the time, with one day a month reserved for preventive maintenance and repairs. The mechanic's objective is now aligned with that of the owner. The incentive structure is now shifted and produces the desired results by "changing the rules of cooperation so that the self-interested rational choices the agent is likely to fulfill the outcomes that the principal desires" [9.2]. Both want the vehicle to be available 98% of the time.

OBCs in the Department of Defense (DoD)

A form of OBCs, known as Performance Based Logistics (PBL), has been the preferred strategy to sustain DoD's systems for over two decades and delivers integrated, affordable performance solutions designed to optimize system readiness. The original intent was to integrate individual stovepipe functions (e.g., procurement; supply; transportation) to improve weapon system readiness but has shifted to balancing system readiness and affordability. This approach has been implemented at the component, subsystem, and system level. There is general agreement that PBL has performed as intended and improved readiness at reduced life-cycle costs, i.e., the cost per unit of performance is lower when a system, subsystem, or component is maintained with a PBL agreement as opposed to a more traditional, transactional maintenance arrangements [9.3].

When OBCs are used to contract for sustainment, the outcome is often a required level of system availability. They are also used in some cases to contract for a performance outcome. This precludes the operator from having to buy the system and pay separately for its sustainment. Some commercial airlines use this approach to contract for engines (see the "Power by the Hour" inset). These arrangements disincentivize the provider's behavior that benefits only themselves but can diminish the quality of the service delivered or the availability of the supported system.

OBCs may offer a resolution to the disconnect that often arises between the motivation for organizations to contract for some services and how they structure those contracts [9.4]. This disconnect occurs when system operators contract for services with the expectation that the superior efficiency of specialized firms will deliver those services more cheaply and reliably. However, when rigid process specifications are used (which detail how the work is to be performed), the operators hinder the provider's ability to innovate and thereby minimize or negate the provider's primary advantage. The strictness of these contracts does not allow a provider to exploit innovations and the resulting efficiencies. Such a contractual arrangement is, in part, self-defeating. By contracting for performance specifications, OBCs enable these sustainment providers to profit from reduced costs or innovation. Furthermore, reducing the focus on specifications also decreases expenditures by reducing the need for oversight by the operator's organization.

"Power by the Hour"

Commercial airlines were among the first to embrace OBCs. Performance-based contracting in this industry took the form of "power-by-the-hour" (a Rolls Royce registered trademark) contract. With this approach Rolls Royce provides aircraft engines, along with the required maintenance. The airline pays a fixed fee per flight-hour that the engine is in use instead of paying a fee for the engine maintenance [9.5].

This approach was initially offered in 1962 to support the engine on the Hawker Siddeley 125 business jet, when a complete engine and accessory replacement service was offered on a fixed-cost-per-flying-hour basis. This aligned the interests of the engine manufacturer and aircraft operator; the operator only paid for engines that performed [9.5].

(Continued)

> (*Continued*)
>
> As their program matured, additional features were offered
> that included Engine Health Monitoring, which tracks on-
> wing performance using onboard sensors; access to leased
> engine to replace an engine during off-wing maintenance; and a
> global network of authorized maintenance centers. This service
> allows operators to remove risk related to unscheduled main-
> tenance events and make maintenance costs planned and pre-
> dictable [9.5].
>
> GE offers a similar model for aircraft engine and has also
> adopted it to other industry sectors like power generation
> [9.6].

As a result, the approach that operators take to provide sustainment for their systems create powerful incentives and disincentives
for the sustainment provider. Consequently, when using the legacy
transactional approach, sustainment providers have less incentive to
perform preventive maintenance since they stand to gain from more
lucrative repairs in the future.

For example, research has found that General Motors' greatest
return on capital comes from after-sales maintenance; it was significantly higher than that earned through the sale of its cars [9.7]. One
could argue that this may have reduced the incentives to manufacture reliable cars. In contrast, with OBCs, such as the fixed sum per
flight-hour schemes used with commercial airlines, the sustainment
providers are only paid for the hours the engine is in use, thereby
incentivizing behaviors that ensure the engine has greater availability
at lower cost.

The principles of OBCs have also led to changes in other industry
sectors, such as health care, where it is known as pay for performance. With pay for performance financial incentives are introduced
to achieve optimal patient outcomes, instead of compensating medical personnel strictly for services performed. The similarities to OBCs
are clear. Furthermore, the clear links between the private sector
health care industry and the public sector health care demonstrate
that OBC concepts work both within industry sectors as well as

between them. The Centers for Medicare and Medicaid Services supports a Value-Based Purchasing system, which will pay "for inpatient acute care services based on the quality of care, not just quantity of the services they provide" [9.8].

9.3 Contract Types

The federal government, an early adopter of OBC, uses two basic contract types; these are fixed-price and cost-reimbursable contracts. These are described in detail in the Federal Acquisition Regulation (FAR), the primary set of rules regulating federal acquisitions. These contract types have some fundamental differences (see Table 9.1); both contract types are used to implement OBCs.

Within these two basic types there is a range of variations. At one end of the contractual spectrum is the firm fixed-price contract, under which the contractor is fully responsible for all of the performance costs and enjoys (or suffers) the resulting profits (or losses). At the other end of the spectrum is the cost-plus fixed-fee contract, in which all allowable and allocable costs are reimbursed, and the negotiated fee (profit) is fixed.

In between the extremes described above are various contract types that can include additional incentives and awards. Decisions about the appropriate type of contract to use are closely tied to the system operator's needs and can go a long way to motivating superior performance – or contributing to poor performance and results. Market research, informed business decisions, and negotiations will determine the best contract type to use for each specific application.

Firm-fixed price (FFP) contracts are generally considered the preferred option by the customer. With this type of contract, there is

Table 9.1.   Contract category characteristics.

	Cost Reimbursement	Fixed-price
Promise	Best effort	Shall deliver
Risk to supplier	Low	High
Risk to customer	High	Low
Cash flow	As costs are incurred	On delivery
Administration	Maximum customer involvement	Minimum customer involvement
Fee/profit	Fee	Profit

a predetermined value that does not change throughout the period of performance. The contractor is required to deliver a product or provide services regardless of its actual costs. Full responsibility for all costs, and the resulting profit or loss rests with the provider, irrespective of the time spent on the job or materials purchased. As a result, most of the risk is shifted to the seller.

Firm Fixed-price (FFP) Contracts can be High Risk

Some risks with FFP contracts that may be hard to anticipate can significantly change the supplier's costs since FFP contracts do not allow any provision for economic adjustments. Consequently, if some of the assumptions in the contractor's cost estimate for a long-term OBC (e.g., inflation or changes in currency exchange rates) turn out to be wrong, the contractor can face significant losses since they have contracted to deliver the agreed-upon service, at the agreed-to price.

For long-term OBCs where, for example, inflation may vary significantly, a more appropriate contract type would be a fixed-price with economic price adjustment contract (FP-EPA). This type of contract is similar to an FFP, but it provides for an adjustment to the price, where both the supplier and customer agree to predetermined criteria to adjust the price based on market conditions that are out of control of either party. This approach would reduce risk to the supplier, which otherwise would have to be included in the contract resulting in a higher price.

This contract type is suitable for situations where both the buyer and seller have a clear understanding of the scope of work and are confident of the cost of the item or service to be delivered. They provide an inherent incentive for the seller to meet the performance requirements while controlling costs to increase their profits.

However, these contracts may not provide the best solution for all system sustainment requirements. "Force fitting" a firm fixed-price contract type, when there is not enough certainty in the data to accurately project the potential sustainment requirements, can result in much higher prices. In those cases, contractors will strive to ensure the price is high enough to cover the worst-case risks. A firm fixed-price contract, which best utilizes the primary profit motive of the business enterprise, should only be used when the risk involved is minimal or can be projected with acceptable confidence. However, when a reasonable basis for firm pricing does not exist, other contract types should be considered. Negotiations should strive to select a contract type (or a combination of types) that will link the provider's profit to their performance.

The other primary contract type used by the federal government is the cost-reimbursement contract. This contract provides for payment of allowable incurred costs to the extent that is prescribed in the contract. With this contract type, a cost ceiling is established that the contractor exceeds at their own risk unless approved by the government. Consequently, the contractor has minimal responsibility for or incentive to control costs or improve performance. These contracts are used when the uncertainties involved do not permit costs to be estimated accurately enough to use a fixed-price contract, and, as a result, the government assumes most of the risk. Cost reimbursement contracts can also include incentives.

Additional incentives, either positive, negative, or a combination of both, can be used with both fixed price and cost-reimbursable contracts to encourage better performance for specific achievements. Positive incentives provide the opportunity for the provider to realize increased profit (incentive fee and/or award fee) and increase the contract term (award term). The performance incentives should be challenging but achievable. The objective is to incentivize providers to strive for outstanding performance; however, the providers should not be penalized when they just meet the performance requirements. When negative incentives are used, the penalty should be structured to correlate to the value of the service that has been lost or impacted when outcomes fall short of the customers' desired levels.

The federal government uses several different types of fixed-price and cost reimbursable contracts to accommodate different scenarios, enabling government agencies to customize their contracts for specific situations. Perhaps the most used is the fixed-price incentive (firm target) contract. With this contract type, the contractor's profit is linked to actual performance, and as a result, the contractor is incentivized to control costs. This contract type stipulates a target cost and a target profit. The contract also specifies a price ceiling (the maximum that may be paid to the contractor) and a profit adjustment formula. This formula is referred to as a share ratio and is intended to incentivize the contractor to control costs. There is

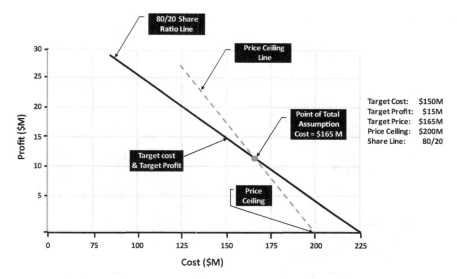

Fig. 9.1. Fixed-price incentive contract profit adjustment formula.

no profit ceiling or floor, however. These elements are all negotiated and can result in more or less risk for either the government or the contractor.

To illustrate, let's examine a specific notional example; the program has a target cost of $150 million, with a target profit of 10% or $15 million. The resultant target price is $165 million. The price ceiling is $200 million, and the profit adjustment formula (share ratio) is 80/20. This formula rewards the contractor with more profit if final costs are less than the target cost. In this case, the contractor would receive 20% of the savings below the target cost. The price ceiling of $200 Million is represented by a dashed line in Figure 9.1. The final contract price and profit are calculated based on the final negotiated cost.

Another type of contract is one with award fees. This contract type is used when contractor performance cannot be measured objectively, making other incentives inappropriate. The award is generally payment of an additional fee but can also come in the form of an "Award Term". This performance incentive ties the length of a contract's term to the contractor's performance, i.e., the contract term can be extended for "good" performance or reduced for "poor" performance. These contract extensions are seen as an attractive incentive to commercial support providers because they can provide stability to projected work orders and add to the firm shareholder's value (see the JSTARS inset).

JSTARS Award Term

The U.S. Air Force's Joint Surveillance Target Attack Radar System (JSTARS) is an airborne command, control, intelligence, surveillance, and reconnaissance platform used for air to ground battle management and surveillance operations. The aircraft can detect, locate, classify, track, and target hostile ground/surface vehicle movements and relay real-time information through secure data links [9.9].

In 2000, the Air Force awarded a PBL contract to Northrop Grumman. The Air Force's approach was to use the contract structure to develop and maintain an atmosphere of cooperation between Northrop Grumman and the Air Force and to incentivize long-term contractor investment in product support improvements. To accomplish this, the Air Force used a two-pronged strategy; they used award fee incentives to encourage short-term performance, and they used award-term incentives simultaneously to promote long-term performance. The resulting contract structure is cost-plus award fee with award terms. The basic contract included a base of 6 years with 16 possible option years for a total potential contract term of 22 years [9.9].

The contractor received a DoD award for their performance on this program: the contractor achieved a 96% effectiveness rate, and a Readiness Spares Packages (RSPs) fill rate above 96%, both exceeding the performance requirement, as well as achieving a reduction in the ownership cost [9.10].

The JSTARS PBL contract has proved effective since it incentivizes the contractor to exceed performance requirements as well as achieve cost reductions. Based on their performance, Northrop Grumman was awarded additional award terms. As of the end of 2010, Northrop Grumman had already earned contract term extensions through 2018 [9.10]. The most current contract was signed in January 2020 [9.11].

It is essential to structure the contract to provide meaningful incentives and not include incentives that the contractor will be unable to achieve or will achieve regardless of the existence of an incentive. Negative incentives are generally included as a complement to the reward. The positive and negative incentives in an incentive-type contract are specified as a ratio.

OBCs can and should evolve along a trajectory over a system's life. With new systems, when data may be scarce, and uncertainties are significant, cost-plus reimbursement contracts enable the collection of sufficient data by both the customer and the service provider to develop a cost baseline. Once the system failure modes and rates and the resultant costs have stabilized, the program should transition to a fixed-price contract. Now the providers will be paid a fixed fee or rate (e.g., per hour, per mile) as long as the required system availability is maintained. As the contractor takes on more risk, reflected in the type of contract used, the contractor's opportunity to increase profit also increases.

Over time, the provider can also make improvements to its supply chain, logistics networks, operations, and the system itself to reduce costs and increase profitability. In the "latter stages" of its evolution, the goal should be to transition to a firm fixed-price contract to continue to incentivize the provider to make performance improvements and reduce costs.

As a system approaches retirement, a shift to a cost-plus contract may allow the operators to, once again, better balance costs, risk, and performance requirements. One final point, the decision on contract type is not necessarily an either-or. Hybrid contracts, those with both fixed-price and cost-type tasks, can be written and are also common.

9.4   Implementation of Outcome-Based Contracts

Outcome based contracting, when used appropriately, can reduce sustainment costs relative to traditional, transactional approaches. When used for sustainment, OBCs generally transfer inventory management, technical support, and the supply-chain functions to a provider who guarantees a level of performance at the same, or reduced, cost. Instead of buying spares, repairs, tools, and data in individual transactions, the customer purchases a predetermined level of outcome (e.g., availability) to meet the customer's objectives.

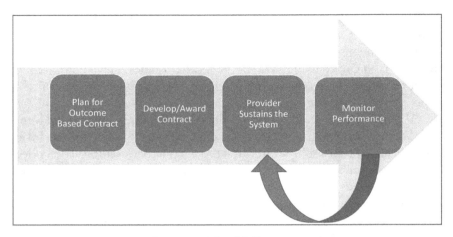

Fig. 9.2.   The OBC process.

The optimal OBC contract is a multi-year agreement wherein the user purchases sustainment in an integrated way, to include elements of the system's supply chain. Long-term contracts allow the provider to incur up-front investment costs in the beginning stages of an OBC contract that are later offset by future cost avoidance. As previously stated, traditional sustainment contracts incentivize the provider to sell parts, while an OBC approach aligns the objectives of the service provider with those of the customer; and motivates the provider to reduce failures and resource consumption. OBCs will generally involve more planning and a greater understanding of the supplier market to develop effective contracts. Figure 9.2 summarizes the OBC process.

A critical step in contracting for sustainment, especially with OBCs, is to work with all the stakeholders and identify what constitutes success for the sustainment effort. That is, what are the requirements for the system performance. In most cases, these requirements will be some form of availability and reliability allocated to the system, subsystem, and/or component level.

Once the performance requirements are identified, the system's current level of performance should be determined to establish the baseline against which future performance can be measured. You need to know where you started, to know how far you've progressed. The analysis of the "as-is" sustainment strategy and contract structures helps to determine if a change to an OBC approach is warranted. If the result indicates a shift to an OBC arrangement will successfully improve the system's performance, while reducing life-cycle costs, then a more detailed analysis of alternative OBC strategies may be indicated. When developing a sustainment strategy for a new system, the initial sustainment baseline may need to be based on engineering design data and models since data from operational systems may not be available.

Contract metrics should be developed to complete the planning phase (there is a more detailed discussion of metrics in the next section). These metrics should be derived from the top-level sustainment requirements of the system, which generally include a measure of availability and reliability. These system-level metrics can then be decomposed and allocated to the subsystem and component levels to ensure that the system level requirement is met. The sustainment provider must have authority and control of the factors that affect system availability and, or reliability within their area of responsibility. Program sustainment metrics should always consider the cost of collecting the necessary data; costly data collection processes should only be used for critical system elements.

In some cases, it may be appropriate to ask the sustainment provider to propose metrics since they may have the best insight. For some program attributes, such as quality, existing commercial quality standards, such as International Standards Organization (ISO) 9000[1] should be considered.

In preparation for developing the contract, the system operator should conduct a business case analysis (BCA) to identify and compare system sustainment alternatives to assess their potential results and assess their impacts on operations, costs, risks, and their associated sensitivities. The examination of sustainment options could also include an analysis of performance, reliability, and availability options, evaluating their impact on cost, benefits, and risks. The result of the analysis should identify the cost and benefits for the

---

[1]ISO has established the ISO 9000 series used by U.S. firms to identify suppliers who meet the quality standards. The term "ISO 9001 2000" refers to a set of new quality management standards that apply to all types of organizations in all kinds of areas.

alternative solutions and, consequently, the approach that meets all the system sustainment requirements at the lowest support costs. Based on the results of the BCA, a sustainment approach can be selected; then the chosen approach can be used to develop a performance work statement and solicit and identify a service provider. Once a service provider is selected, the customer can review and help structure the OBC. The contract can then be awarded.

The period of performance for OBCs should be long enough to enable the contractor to make and recover investments in product and process improvements. If the contracts are too short, the provider will not make these investments since they will be unable to recover their investment. The Ministry of Defense in the United Kingdom, for example, awards 30-year outcome-based sustainment contracts that have reviews and exit ramps every 5 years. Consequently, these contracts will be competed less frequently [9.12]. This approach may conflict with policy and guidance from many organizations that mistakenly emphasize frequent competition as the only approach to control costs.

Recent research has explored a more deliberative structure, particularly as it applies to a population of fleet of systems, such as a fleet of airplanes or a wind farm. This methodology, known as "contract engineering", strives to integrate the system's engineering design in conjunction with the contract design [9.13].

Contract engineering uses dynamic simulation to provide a more accurate and realistic estimate of the system's life-cycle costs by studying contractual and performance parameters in an integrated cost-performance model. The objective of contract engineering is to provide the acquisition managers the information they need to assess and compare a variety of contract structures, to select the best structure, and finally to price and negotiate this contract. From the viewpoint of a system designer, the process of designing contractual terms that examine the performance metrics, incentives, the payment model, uncertainties, and performance assessment, represents a multidisciplinary design process that should be incorporated into the broader design process of complex systems and their sustainment [9.13].

System sustainment programs will need to establish periodic reviews to examine performance and costs to provide feedback on the effectiveness of the sustainment program. Evolving system requirements or system design changes may require corrective measures. When problem areas are identified, remedial actions can be identified. An important question, especially for the reviews early in the period of performance of an OBC, is to ask, "Are we measuring the right things?" OBCs should contain provisions for review of metrics and measures and enable changes when required.

9.5 Performance Metrics

A successful OBC will depend on performance metrics that are straightforward, measurable, and achievable. The contracts will specify performance metrics that will link the sustainment provider performance to the requirements of the system's *users*. These will generally be related to the system's life-cycle cost and readiness to perform its function [9.14]. In this context, a metric is a verifiable quantitative measure relative to a reference point and that is consistent with how the provider delivers performance. The metric should be based on an agreed upon set of verifiable data, along with a documented and agreed to process for converting this data into the measure. The metric is then compared to a reference (an absolute standard or an internally or externally developed standard) that acts as a basis. Finally, to be effective, metrics should be expressed in meaningful terms so that they make sense to the person using the metrics [9.15]. See Table 9.2 for some example metrics [9.16].

Table 9.2.   Example metrics (adopted from the DoD's PBL Guidebook).

Name	Description
Operational Availability (A_o)	The fraction of time that a system or group of systems are operationally capable of performing their required function
Reliability (R)	The probability that the system will perform without failure over a specified interval under specified conditions
Mean Time Between Failure ($MTBF$)	For a particular interval, the total functional life of a population of an item divided by the total number of failures (requiring corrective maintenance actions) within the population
Turnaround Time (TAT)	The amount of time elapsed between when an action is initiated and its completion (could apply to maintenance, repair, logistics, etc.)

Developing a few top-level metrics that can quantitatively measure how well the sustainment provider delivers the required outcomes and incorporating them into the contract is key to developing an effective incentive structure. When using OBCs for less than comprehensive sustainment of a system (e.g., for a component or subsystem), it is nonetheless critical to have integrated system models to monitor the system level operational and support BCA [9.15]. Developing a sound outcome metric strategy is critical for the OBC's success. An appropriate OBC contract structure aligns the customer's objectives with those of the support provider, leading to a win-win scenario. Conversely an inappropriate structure can create perverse incentives and result in undesired or unintended consequences.

The metrics should be tailored to the user's operational requirements and synchronized with the scope of responsibility of the sustainment provider. The user will use these metrics to evaluate program performance and determine compliance with the OBC, as well as determine how the sustainment providers will be rewarded or penalized. Other data and metrics will also generally be collected; the metrics structure is generally hierarchical with some measures that can help serve as a diagnostic for the higher-level metrics (e.g., to help research the rates and causes of failures). An effective metrics construct will ensure the sustainment provider's activities and objectives are aligned with the system's users so that they help to shape performance without obstructing it, i.e., the metrics should require outcomes without specifying processes. In this way, the program can maximize the opportunities for the sustainment provider to innovate to improve performance and increase efficiencies.

Metrics should be selected or constructed to encourage performance improvement, effectiveness, efficiency, and innovation. There is no perfect metric but selecting an inappropriate contract type with a metric that is too simple can create perverse incentives and lead to undesirable results. See the Stryker case inset. Choosing an appropriate complementary set of metrics will promote the desired behaviors and outcomes while minimizing unintended consequences.

The U.S. Army High Mobility Artillery Rocket System (HIMARS) Program

HIMARS is a U.S. Army wheeled rocket and guided missile launcher for which logistics and sustainment relies on a type of OBC known as PBLs within the U.S. Department of Defense. The HIMARS program uses metrics to track performance in the following areas: system readiness status, average response time for critical non-mission capable launcher failures, average repair time in the field, and average depot repair turnaround time. Since its inception, the HIMARS OBC has consistently met or exceeded cost and performance objectives, having twice received the Secretary of Defense Performance-Based Logistics Award, which recognizes government and industry teams that provide the DoD with superior operational capability – the program achieved a 98.7% availability [9.17].

Inappropriate Metrics can Create Perverse Incentives U.S. Army's Stryker PBLs

The M1126 Stryker is a rapidly deployable wheeled armored vehicle. The initial PBL for the Stryker vehicle was awarded to General Dynamics Land Systems (GDLS) in 2000 as part of a cost reimbursable larger contract for vehicle manufacture and delivery.

The cost-reimbursable contract was chosen to provide maximum flexibility to meet the rapidly changing conditions, and at the same time enabling U.S. Army officials to collect sufficient data associated with different performance levels so that the program could transition to a Firm Fixed-price (FFP) contract at a later point. The program used the system's operational readiness rate (ORR) as the single system metric, which was set at 98% ORR for fielding and training exercises and at least a 90% ORR for deployed vehicles. The system averaged a 96% ORR when deployed to Iraq in the 2003–2006 timeframe.

(*Continued*)

(Continued)

The U.S. Army also consistently noted that contractors were providing impressive levels of support and were generally more knowledgeable and efficient performing Stryker maintenance than their military counterparts.

However, when considering cost, performance is less clear. A 2012 DoD Inspector General Report concluded that the program's use of the single metric, readiness, in conjunction with a high-ceiling, cost-plus contract, created a perverse incentive for the contractor to accumulate significant excess inventory to meet the readiness metric. Perhaps the U.S. Army could have controlled costs better by tying a fixed fee to an agreed-upon cost-per-mile metric [9.18].

9.6   Modeling Contracts for Availability

Contracts can be modeled from several viewpoints; however, it is most useful to model the class of contracts discussed in this chapter from the contractor's point of view as a payment model that addresses the desired outcome constraints in which performance and cost are weighted into a single factor to simplify the payment rationale [9.19],

$$R_v = \omega + \alpha C + vA \qquad (9.1)$$

where R_v is revenue received by the contractor, C is the cost to the contractor, A is the availability, and ω, α and v are the contract parameters chosen by the customer and described in the contractual document. ω is the fixed payment. α is the customers' share of the contractor's costs of operation and v is the penalty or reward rate for achieved availability below or above the contractually required level.

By varying the contract parameters in Equation (9.1), different classes of cost-driven and OBCs can be modeled. For example, $v = 0$ and $\alpha = 0$ is a firm fixed-price contract, and $v = 0$ and $\alpha = 1$ is a cost-plus contract with full reimbursement. Since this model is completely known (transparent) to the contractor it is safe to assume that they optimize their decisions based on the above model. However, their decisions might incur costs on the customer that are outside of the scope of the contract, for example maintenance costs the customer incurs after the contract is over.

From the viewpoint of the customer, these contracts can be modeled by a Stackleberg game[2] in which, depending on the contract designed by the customer, the contractor will optimize its strategy. From this point of view the contract can be modeled by a two-level optimization problem [9.20], which is beyond the scope of this text, however, in this section, we will introduce a simple spares provisioning model under a PBL OBC that uses Equation (9.1).

Availability contracts that reward achieving specific availability targets cause a shift from cost minimization to profit maximization. Profit maximization becomes the goal that is leveraged to form OBCs between the public and the private sectors. The inventory management problem changes into a profit maximization to contractors driven by the availability achieved (and/or other outcomes) problem. The optimization problem for the contractor becomes, in part, an optimization of the spares quantity and location that maximizes the contractor's profit when costs are minimized, but the availability achieved by the customer is rewarded or penalized.

The analysis of incentives includes an agent who is responsible for producing an output (which could be availability) and a principal who owns the output. The principal and the agent sign a contract and the agent receives compensation as specified in the contract. The modeling of this type of arrangement is called agency theory. Equation (9.1) represents a linear contract function where vA is the performance-based incentive (other functional forms for contract functions exist, e.g., exponential).

One way to define an OBC is to define the contractor's revenue as shown in Figure 9.3. In Figure 9.3, the zones are defined as [9.21, 9.22]:

- *None Zone*: In this zone the contractor receives no compensation.
- *Penalty Zone*: In this zone the contractor will receive compensation, but a penalty is assessed.

---

[2] A Stackleberg game (also known as a leader-follower game) is a strategic game in economics in which at least one player is defined as a leader who makes a decision and commits to a strategy before other players (who are the followers) make their decisions sequentially thereafter.

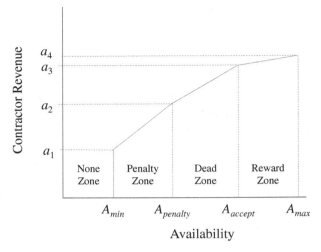

Fig. 9.3.   Revenue model, after [9.21, 9.22].

- *Dead Zone*: In this zone the contractor receives either revenue that linearly increases with availability, or possibly constant revenue.
- *Reward Zone*: A cap on revenue may be defined in this zone.

Following [9.21] the revenue model (based on Equation (9.1) and Figure 9.3) that corresponds to the zones defined in Fig. 9.3 is given,

$$
R_v = \begin{cases}
0, \ if \ A < A_{min} \\
a_1 + v_1(A - A_{\min}), \ if\, A_{\min} \leq A < A_{penalty} \\
a_2 + v_2(A - A_{penalty}), \ if\, A_{penalty} \leq A < A_{accept} \\
a_3 + v_3(A - A_{accept}), \ if\, A_{accept} \leq A < A_{\max}
\end{cases}
\qquad (9.2)
$$

Where the slopes are given by,

$$
\begin{aligned}
v_1 &= \frac{a_2 - a_1}{A_{penalty} - A_{\min}} \\
v_2 &= \frac{a_3 - a_2}{A_{accept} - A_{penalty}} \\
v_3 &= \frac{a_4 - a_3}{A_{\max} - A_{accept}}
\end{aligned}
\qquad (9.3)
$$

Using the availability given by Equation (4.13) with $n = 1$,

$$A = \sum_{i=0}^{k} \frac{(\lambda t)^i}{i!} e^{-\lambda t} \tag{9.4}$$

then the profit for k spares is the revenue minus cost given by

$$P_k = a + v \sum_{i=0}^{k} \frac{(\lambda t)^i}{i!} e^{-\lambda t} - kC_P \tag{9.5}$$

where the last term in Equation (9.5) is the cost of purchasing k spares (C_P is the purchase price of the spare part). Equation (9.5) ignores cost of money and holding costs for the spares. The marginal increase in profit for increasing the number of spares by one is given by,

$$P_k - P_{k-1} = v \frac{(\lambda t)^k}{k!} e^{-\lambda t} - C_P \tag{9.6}$$

Thus, one additional spare will result in a profit when $P_k - P_{k-1} > 0$ or,

$$\frac{(\lambda t)^k}{k!} e^{-\lambda t} > \frac{C_P}{v} \tag{9.7}$$

The model above, and Equation (9.7) provides the basis for determining the spare quantities within a multi-echelon inventory system that maximize contractor profit under an availability contract. A solution for a multi-item, multi-echelon inventory system is given in [9.21].

As a simple example, assume that we are operating in only the Dead Zone. How many spares should be carried in order to maximize the contractor's profit? The assumed input data is:

$$a_2 = \$1M$$

$$a_3 = \$2.6M$$

$$A_{penalty} = 0.95$$

$$A_{accept} = 0.98$$

$$\lambda = 0.42 \, \text{failures/year}$$

$$t = 5 \, \text{years}$$

$$C_P = \$4.3M$$

The profit (P) as a function of k from Equation (9.5) is shown in Figure 9.4 for this case.

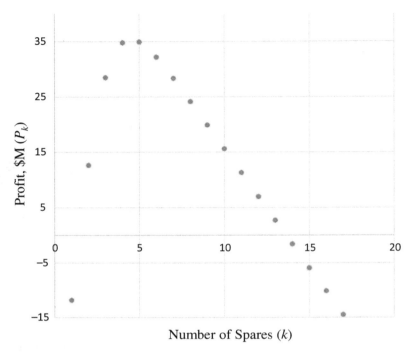

Fig. 9.4.   Profit as a function of number of spares for a single band model.

For this case, the optimum number of spares is 5. We could have also determined this by solving Equation (9.7) for k (in this case, $k = 4$).

Consider a more complex example that spans all the bands,

$$a_2 = \$200K$$

$$a_2 = \$1M$$

$$a_3 = \$2M$$

$$a_4 = \$2.5M \text{ or } \$4M$$

$$A_{min} = 0.8$$

$$A_{penalty} = 0.9$$

$$A_{accept} = 0.96$$

$$A_{max} = 1$$

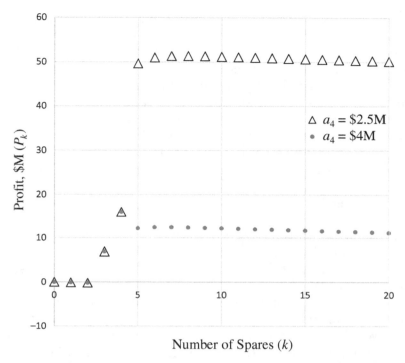

Fig. 9.5. Profit as a function of number of spares for a multi-band model.

$$\lambda = 0.42 \, \text{failures/year}$$
$$t = 5 \, \text{years}$$
$$C_P = \$100\text{K}$$

Figure 9.5 shows the results of the multi-band model. In this case if a_4 is \$2.5M, the optimum number of spares for the contractor to hold is 7, while if a_4 is \$4M, profit is maximized with holding 4 spares.

9.7 Intellectual Property and Technical Data Package Ownership

An important factor when contracting for sustainment is the need to consider intellectual property (IP) and technical data package (TDP) ownership.

Throughout much of the last century most engineering and manufacturing activities relied on hardcopy documents, including two-dimensional drawings to communicate engineering data. Beginning in the last quarter of the 20th century, the information revolution initiated a significant paradigm shift. As information technology improved, product definition and specification could be represented in three-dimensional computer models; these models ushered in a revolution in how products are developed and manufactured. Now these models could be overlaid with metadata that could be used to help automate the manufacturing and inspection processes. As discussed in Section 2.5, these models, along with other supporting data form the basis of a TDP[3] needed to manufacture and maintain the product. As new manufacturing technologies are adopted, e.g., additive manufacturing, much of a firm's value resides within these TDPs, with their associated IP (see Section 10.1).

When considering the sustainment of long-lived critical systems, the total ownership costs of TDPs, associated IP and their licensing rights required for the operation, maintenance, modernization upgrades, and sustainment needs to remain competitive and affordable over the entire product life cycle of key programs. The underlying assumption is that the cost of the management of this digital data will be offset by future benefits, either using the data directly or through the introduction of competition.

Making sound judgements about life-cycle data needs, evaluating the existing and new business cases, negotiating, and contracting for the data rights, and making the relevant program data accessible for use over an extended period while maintaining its currency are challenging tasks. Additionally, program managers will need to account for the possibility that data needs may change over the life cycle of

---

[3]Military Standard 31000B defines a TDP as the authoritative technical description of an item. This technical description supports the acquisition, production, inspection, engineering, and logistics support of the item. The description defines the required design configuration and/or performance requirements, and procedures required to ensure adequacy of item performance. It consists of applicable technical data such as models, engineering design data, associated lists, specifications, standards, performance requirements, quality assurance provisions, software documentation and packaging details.

the project as the program matures. Early product support, as well as support through long-term OBCs, may be provided by the OEM and would not require IP ownership. However, as the program matures, support needs and requirements may change if the system operator assumes a greater role in the systems sustainment. Consequently, the system operator may need greater access to, or ownership of, relevant technical data to maintain a competitive market and preclude vendor lock-in.

With many unknowns and uncertainty in program life cycles, this issue will require careful analysis by the acquisitions professionals to avoid over-authorization of data access or rights, with the associated increases in the costs of acquiring the rights to support the system's sustainment.

Consequently, system operators, in conjunction with their sustainment providers, should develop a robust IP strategy that identifies and manages the full scope of IP related matters, including technical data, computer software deliverables, patented technologies, and appropriate licensing rights. This strategy should consider the entire product life cycle, starting with inclusion in the initial acquisition strategy and program planning decisions, through operation and sustainment.

Finally, despite the potential benefits of owning a program's IP and data rights, these data are generally not static. TDPs are often changed to improve component's performance, reliability, maintainability, as well as to improve production processes. Recognizing this dynamic nature, OEM contractors that designed and developed the design, as well as managed the changes and improvements, are often in the best position to maintain, update, and modernize the TDP, and ensure integration with the system or sub-system.

The current goal of many programs, especially with federal government programs, is to negotiate for and acquire a complete TDP in anticipation of future needs, even though these may be unspecified and uncertain. Therefore, acquiring TDPs by default, in some cases, risks paying for TDPs that may be unneeded. An approach that can help deal with this program uncertainty is to use a real options approach to the acquisition of TDP's and associated IP (see Section 10.4.1).

9.8 The Benefits and Challenges of Outcome-based Contracts

9.8.1 *Benefits*

When implemented, OBC shifts the focus of the *customer*'s efforts from transactions to identifying performance outcomes and assigning responsibilities. The objective is to develop accountability, instead of relying on control. With OBC, active management of the sustainment process (e.g., forecasting demand, maintaining inventory, and scheduling repairs) becomes the responsibility of the support provider. Traditional system sustainment dictates processes and design specifications, which has the effect of constraining innovation and process improvement. Sustainment providers and equipment manufacturers are incentivized to sell more repair parts, as opposed to developing and implementing reliability improvements. OBCs change the incentives for the supplier; the provider is now incentivized to improve the reliability of their systems and reduce inventories of spare parts, to increase profit.

The U.S. Department of Defense, an early adopter of OBC has abundant empirical data that demonstrates that OBCs, when properly implemented, produces desired outcomes in the key performance areas of availability, reliability, logistics footprint, and cost.

Major DoD weapon system programs, e.g., the C-17 (a cargo aircraft) and the F/A-18 (a fighter aircraft), have reduced their sustainment costs by hundreds of millions of dollars, while other systems and subsystems, including F-22 (a fighter aircraft), UH-60 (a utility helicopter) avionics, and the F-404 jet engine (used in front line fighters and trainers) have seen dramatic improvements in availability and cycle times (i.e., logistics response and repair turnaround) [9.23].

Empirical analysis has also demonstrated that OBC contracts incentivize reliability improvements of 10–25%, compared to more traditional transactional time and material contract approaches [9.24]. Other government reports (e.g., [9.25]) have concluded that OBCs offer distinct benefits that are difficult to achieve using traditional transactional approaches.

To summarize, there are three distinct advantages associated with the use of OBC contracting:

- OBCs identify outcome performance goals and ensure program responsibilities are assigned. The objective of OBCs is to buy measurable outcomes based on the *customer*'s system performance requirements. These requirements should be high level, and include only a few simple, realistic, consistent, and easily quantifiable metrics. When properly developed, OBC metrics clearly define the *supplier*'s responsibilities.

- OBCs provide incentives for improving system performance and attaining performance objectives. When properly structured, the OBC will align the objective of the *supplier* with that of the *customer* and will incentivize *suppliers* to assume greater responsibility for providing ongoing improvements to their system and products, by improving designs and processes and implementing best practices.
- OBCs can reduces the cost of ownership. The reduction results from the decline in inventories, improved supply chain efficiency, replacement of low-reliability components, and increased system availability.

9.8.2 *Exploring the challenges with OBCs*

Although OBCs can provide many program benefits, successfully implementing them is not without challenges. First, they are more difficult to develop. Two principal elements of OBC that will directly impact their success are the metrics and incentives. Establishing appropriate metrics is critical to the success of these arrangements, because if metrics are inadequately developed or inconsistently implemented, measuring the provider's performance in achieving the operator's objectives will be more difficult. Therefore, inherent in successful OBC is the challenge of developing appropriate, effective performance metrics. The incentive structure must also be adequate to motivate the desired behaviors, with clear guidelines when the attribution of performance improvement initiatives includes contributions from the system's operators. Inappropriate metrics and incentives may result in perverse behavior and not achieve the desired results (as highlighted in the Stryker case).

Second, OBCs can be more challenging to develop and manage than the more traditional transactional contracts and require a different skillset. Specifically, the operator's organization workforce often does not have a thorough understanding of how to structure contracts with the appropriate the incentives, penalties, and the contract types to motivate sustainment providers to provide superior support, while reducing costs. Accordingly, they must be trained in the appropriate use of OBCs, and how to structure them with suitable metrics and incentives to achieve program objectives.

An appropriate OBC uses its structure and incentives to align the objectives of the *customer*, with those of the support provider, leading to a win-win scenario; but as described in the Stryker example, inappropriate structured OBCs can create perverse incentives and result in undesired or unintended consequences. The operator workforce must have a good understanding of what motivates the providers, to ensure that the contractual incentives will achieve the desired outcomes.

OBCs will also require cultural changes within both the customer and supplier organizations. This contract type changes the responsibilities, work processes, and practices of acquisition professionals, since it is dramatically different from the way legacy transactional contracts operate. The customer may perceive lack of control since many decisions will now be made by the sustainment provider. This may produce some uneasiness since the operator is still accountable for the provision of the service.

There are also cultural changes required within the supplier's organization, since their organization, business models, and cost structure may all be optimized for a transactional approach. Additionally, there may also be a lack of continuous improvement processes and culture, which is necessary to full leverage the potential of OBCs.

Finally, OBCs will complicate the relationships with lower tier suppliers. The support provider will now have to decide how to structure these subcontracts, either as transactional or OBC, to optimize the service performance without incurring any excess costs.

References

[9.1] Ng, I. C. L. and Nudurupati, S. S. (2010). Outcome-based service contracts in the defence industry – mitigating the challenges, *Journal of Service Management*, 21(5), pp. 656–674.

[9.2] Taylor, K. and Shaver, M. (2010). Performance-based contracting: Aligning incentives with outcomes to produce results, in *Fostering Accountability*, M. F. Testa and J. Poertner (eds.) (Oxford University Press), pp. 291–327.

[9.3] Boyce, J. and Banghart, A. (2012). Performance based logistics and project proof point – A study in PBL effectiveness, *Defense AT&L: Product Support*, March-April, pp. 26–30.

[9.4] Martin, L. L. (2005). Performance-based contracting for human services: Does it work? *Administration in Social Work*, 29(1), pp. 63–77.

[9.5] Rolls-Royce Media (2012). Rolls-Royce celebrates 50th anniversary of Power-by-the-Hour Available at: https://www.rolls-royce.com/media/press-releases-archive/yr-2012/121030-the-hour.aspx.

[9.6] Kelly-Detwiler, P. (2017). The outcome as a service model is on the rise in the power sector. Available at: https://www.ge.com/power/transform/article.transform.articles.2017.oct.the-outcome-as-a-service-model.

[9.7] Dennis, M. and Kambil, A. (2003). Service management: Building profits after the sale, *Supply Chain Management Review*, 7(1), pp. 42–48.

[9.8] Center for Medicare and Medicaid Services (2021). What are the value-based programs? Available at: https://www.cms.gov/Medicare/Quality-Initiatives-Patient-Assessment-Instruments/Value-Based-Programs/Value-Based-Programs.

[9.9] Northrop Grumman (n.d.). E-8C Joint STARS. Available at: https://www.northropgrumman.com/what-we-do/air/e8c-joint-stars/. Accessed on November 11, 2021.

[9.10] OSD PBL (2010). Office of Secretary of Defense Performance Based Logistics Awards Program for Excellence in Performance-Based Logistics Life Cycle Product Support; PBL Award Package 2011 Systems_JSTARS. 2011. Available at: https://acc.dau.mil/adl/en-US/548825/file/68137/PBL%20Award%20Pkg%202011%20%20System_JSTARS.pdf.

[9.11] Northrop Grumman (2020). Northrop Grumman Continues Joint STARS Sustainment and Modification Work for US Air Force. Available at: https://news.northropgrumman.com/news/releases/northrop-grumman-continues-joint-stars-sustainment-and-modification-work-for-us-air-force.

[9.12] Gansler, J. S., Lucyshyn, W., and Harrington, L. H. (2012). An Analysis of Through-Life Support – Capability Management at the U.K.'s Ministry of Defence, CPPPE Report, June 2012.

[9.13] Sandborn, P., Kashani-Poura, A., Goudarzia, N., and Leia, X. (2017). Outcome-based contracts–towards concurrently designing products and contracts, *Proc. 5th International Conference on Through-life Engineering Services*.

[9.14] Doerr, K., Eaton, D. R., and Lewis, I. A. (2004). Characteristics Of Good Metrics for Performance Based Logistics, *Proceedings of the*

2nd Annual Acquisition Research Symposium of the Naval Postgraduate School, September 30, 2004.

[9.15] Melnyk, S. A., Stewart, D. M., and Swink, M. (2004). Metrics and performance measurement in operations management: Dealing with the metrics maze, *Journal of Operations Management*, 22(3), pp. 209–217.

[9.16] DAU (2016). DoD's PBL Guidebook. U.S. Department of Defense. Available at https://www.dau.edu/pdfviewer?Guidebooks/Performance-Based-Logistics-(PBL)-Guidebook.pdf.

[9.17] Gansler, J. S. and Lucyshyn, W. (2014). HIMARS: A High Performance PBL. The Navy Postgraduate School, August 2014.

[9.18] Lucyshyn, W. and Rigilano, J. (2017). Trends in Performance-Based Services Acquisition, Naval Postgraduate School, July 2017.

[9.19] Kashani Pour, A., Goudarzi, N., Lei, X., and Sandborn, P. (2017). Product-Service Systems Under Availability-Based Contracts: Maintenance Optimization and Concurrent System and Contract Design. in *Advances in Throughlife Engineering Services. Decision Engineering*. Redding, L., Roy, R., and Shaw, A. (eds.) (Springer Nature, Cham Switzerland).

[9.20] Murthy, D. N. P. and Asgharizadeh, E. (1999). Optimal decision making in a maintenance service operation, *European Journal of Operational Research*, 116(2), pp. 259–273.

[9.21] Nowicki, D., Kumar, U. D., Steudel, H. J., and Verma, D. (2008). Spares provisioning under performance-based logistics contract: Profit-centric approach, *Journal of the Operational Research Society*, 59(3), pp. 342–352.

[9.22] Brown, R. E. and Burke, J. J. (2000). Managing the risk of performance based rates, *IEEE Trans Power Syst*, 15(2), pp. 893–898.

[9.23] Lucyshyn, W., Rigilano, J., and Safai, D. (2016). Performance-Based Logistics: Examining the Successes and Challenges when Operating in Stressful Environments, Naval Postgraduate School, July 2016.

[9.24] Guajardo, J. A., Cohen, M. A., Kim, S., and Netessine, S. (2012). Impact of performance-based contracting on product reliability: An empirical analysis, *Management Science*, 58(5), pp. 961–979.

[9.25] OSD (2009). Secretary of Defense Performance Based Logistics Awards Program for Excellence in Performance Based Logistics: Summary of critical accomplishments. Retrieved from DAU.

Problem

9.1 In the multi-band availability contract example in Section 9.6, what is the optimum number of spares to hold if a_4 is \$5M?

Chapter 10

Epilogue – The Future of System Sustainment

As technological innovation continues to accelerate, system sustainment is not standing still as it adapts to the changing landscape. Potentially the defense, political, economic, and supply-chain stars are aligning to place a greater emphasis on system sustainment than ever before.

In this chapter, we briefly describe some of the new technology, policy, and analysis opportunities on the horizon, which will impact system sustainment. Finally, a general outlook on system sustainment is provided.

10.1 Additive Manufacturing for Sustainment

Additive manufacturing (AM) will have a disruptive effect on the development, production, and especially the sustainment of many critical systems. Commonly known as "3D printing", AM is maturing at a rapid pace, and already has a wide variety of industrial applications in a diverse array of manufacturing contexts. The promise of AM is that it enables manufacturing with an economy of scope due to the ability to make an almost limitless variety of parts with the same machine. AM's power lies in its ability to transform three-dimensional digital blueprints, or technical data packages (TDPs), into physical products, without the need for tools or molds. AM offers the potential for significant operational improvements in development (improving rapid prototyping capability); in production (where

AM can also facilitate mass customization); and, finally, in sustainment where AM will enable organizations to print parts on demand, at the time and place of need.

Unlike conventional subtractive manufacturing processes, which rely on milling, drilling, or cutting processes to remove material during the production process, AM is a "printing-like" process that can be used to make three-dimensional objects from computer models, through the addition of layers of materials. With AM, successive layers of melted or partially melted material are bonded together until the final object is formed. AM was first developed in the 1980s, starting with simple photosensitive resins that polymerized when exposed to UV light. Initially, it was only used for prototyping and the production of small objects. As printing technologies, CAD modeling software, and fusing technologies developed, manufacturers have embraced AM in numerous industries [10.1].

While AM techniques have been on the technological horizon for decades, there is good reason to believe that recently AM has progressed to the point where it will have a disruptive effect on manufacturing. In other words, while the technology has been under development for decades, we are only now entering a period where AM can effectively be utilized for the production of critical components in complex systems. This is significant, as industry will now have continuously compounding incentives to further develop and iterate the technology.

It is anticipated, for example, that AM has the potential to enable on-location and on-demand production of selected parts and components, thereby reducing the need for complex supply chains and spare parts inventories. The Department of Defense has demonstrated this concept for simple parts; for example, the 31st Marine Expeditionary Unit based in Okinawa, Japan recently reported the successful test of a printed plastic bumper on one of its F-35B Lighting II aircraft [10.2].

On-demand, on-location, spare part production however is just one narrow slice of AM's value proposition. As the technology is refined, its reliability enhanced, and its production costs further reduced, AM will increasingly become a staple of precision manufacturing. The freedom and flexibility enabled by AM production processes also expands the frontier of possible designs. By being able to form hollow voids in otherwise solid structures, AM enables not just saving on resources but also the production of vastly more complex systems requiring less post-processing. In a similar manner, being able to manipulate the placement of materials with a high degree of precision either eliminates or greatly reduces the need for tooling or molds specific to a given item providing further opportunities for savings [10.3]. This ability to create complex hollow geometries in a single, joint-less and weld-less component is particularly useful. Fewer joints, bolts, and welds when producing a part with AM means manufacturers can both reduce the complexity and weight of a component, while increasing the product's reliability. Manufacturers such as Orbital ATK, have begun to use the technology to produce more complex components for rocket motors.[1] General Electric received the first airworthiness qualification for a AM part for its F110 engine in June 2021 [10.5].

AM, likewise, has the potential to significantly impact system sustainment. In order meet the availability requirement of complex critical systems, organizations often maintain inventories of items for which demand may be highly variable. Long-lead items may be particularly troublesome since the demand must be anticipated and planned for further in advance. The further in advance you must estimate the need, the greater the uncertainty and, as a result, the greater the need for safety stock. AM's potential for distributed on-demand printing capability can reduce the need to maintain these parts in inventory.

AM also shows great promise in extending the lifespan of legacy systems. Major critical systems, e.g., aircraft, are frequently decommissioned prior to the end of their useful life due to depleted supplies of spare parts and the prohibitive expense of restarting/retooling spare part production lines. As long as the TDPs are retained, maintained and available, and feedstock remains available, sustainment service providers will be able to provide support and most importantly spare parts nearly indefinitely. As a result, legacy systems may be economically kept in service much longer than they otherwise would be, due to the ability to "print" spare parts on an as-needed basis.

---

[1]Orbital ATK, announced it has successfully tested a prototype tactical solid rocket motor, which included critical metal components fabricated by additive manufacturing. The printed single piece steel rocket nozzle and closure structure significantly reduced the part count and manufacturing complexity, and is expected to improve system affordability [10.4].

In summary, when AM is used for appropriate components, the potential exists to digitize the supply chain and move physical production forward, shortening the entire process, reducing inventories, and lowering cost as a result. There is also the potential to merge parts, and, as a result, save time and resources when replacing some components due to the elimination of unnecessary assembly steps. Perhaps the most significant benefits of being able to print parts on demand will undoubtedly be an improved ability to more rapidly modernize, refurbish, and repair systems, shortening the sustainment timeframes and improving system availability. However, integrating AM on a large scale will not be as simple as buying the machines and materials and installing them downrange. There are several challenges that must be overcome, these include:

- AM technology and processes will jump start several key shifts in the manner in which producers and manufacturers interact within the digital economy, however, the legacy software and databases that support engineering, manufacturing, and sustainment are generally not interoperable and may not use the same data formats. In order to reach full potential, AM systems will need to be integrated into the organization's IT infrastructure, i.e., enterprise resource planning (ERP) and labor management systems, along with (ideally) AI-assisted CAD programs enabling a greater degree of integration and synchronization of supply chain management and production tasks.
- AM processes will make it easier for a variety of organizations and individuals to manufacture products that were formerly solely produced by manufacturing specialists. However, the current AM capability is far less mature than the traditional manufacturing sector and therefore has much less experience with failure mechanisms, and how parts wear and degrade overtime, while the distributed and digital nature of these processes may make it difficult to identify and attribute the cause of a defect in an AM produced product. Standards and qualification processes for AM need to be developed and adopted to resolve the potential liability issues.
- Finally, the widespread adoption of AM technology represents a marked shift in the production value chain for manufacturers, from value residing within the physical parts and structures built from traditional designs to systems and components produced via AM processes wherein the value resides in the digital technical designs

themselves (in the TDP). Issues involving the management of intellectual property and data rights, along with the heightened need for protection against the exploitation of digital data need to be resolved.

10.2 Sustainable Sustainment

This book has carefully distinguished itself from environmentally-oriented sustainability issues. In this section we address the environmental sustainability of sustainment activities. System sustainment (as addressed in this book) is not only a set of operations and practices that are focused on dealing with system unreliability, it is potentially long-term strategic planning that encompasses all the phases of a system's life cycle including its impacts on social, environmental and economic trends, and its benefits from innovative technologies.

As the sustainment world evolves, sustainment becomes not just the minimization of total life-cycle cost (and/or maximization of availability), but optimizing these targets "without disturbing safety and environ-mental issues" [10.6].

A concept referred to as green maintenance (GM) appeared in the early 1990s [10.6]. With GM, the goal of maintenance is achieved at the lowest cost, least resources, least energy consumption and least waste, i.e., the smallest environment impact. Conventional thinking tends to push systems from preventative maintenance to CBM and PHM (i.e., minimization of the number of maintenance actions, spares and all of their associated infrastructure). However, minimizing environmental impacts requires a careful life-cycle assessment (LCA) of the system sustainment process to determine the actual inventories of resources (materials used, wasted, energy, etc.).

10.3 Digital Twins

Digital twins are virtual representations of real systems. Digital twins are a tool for understanding how an item is performing or will perform against various usage and environmental conditions including monitoring the system's health. A digital twin is a virtual environment in which an item can be analyzed and studied. In many cases the twin will benefit from having actual sensor data from the real system to augment simulations. Although digital twin technology has primarily been associated with creating virtual representations of physical items, it can also be used to analyze and forecast the performance of processes and organizations, e.g., supply chains and sustainment processes. One could envision a world in which every system will have a virtual representation hosted in the cloud that is continuously updated with operational data.

Many references simply conflate system health management (CBM and PHM, and the forecasting of RULs) and digital twins for sustainment. If one is only interested in a digital twin for maintenance planning, this is not unreasonable. However, digital twins can represent considerably more. A digital twin is essentially a virtual supply-chain replica that could consist of hundreds of assets, warehouses, logistics and inventory positions. While there is certainly value in simply collecting and archiving current and historical data for analytics, the real power of the digital twin is realized when it covers an asset's entire life cycle and is used to forecast supply-chain disruptions and compromises (timing and duration).

While the concept of digital twins for supply chains has been embraced in the commercial product world, it has yet to find significant traction in the sustainment supply-chain space.

10.4 Options-based System Sustainment

In the financial world, options are derivative financial instruments that specify a contract between two parties for a future transaction on an asset at a reference price. For financial options the buyer of the option obtains the right, but not the obligation, to engage in the specified transaction at a specified future date. The seller incurs the corresponding obligation to fulfill the specified transaction at that future date. An option that conveys the right to buy something at a specific price is called a "call" option; an option that conveys the right to sell something at a specific price is called a "put" option.

The version of financial derivatives that is applied to real assets is called real options.[2] Real option takes a different perspective than

---

[2]The term "real options" was originated by Stewart Myers at MIT in 1977 [10.7]. Myers used financial option pricing theory to value non-financial or "real" investments in physical assets and intellectual property.

DCF (see Appendix A) on the valuation of cash flows. Real options analysis is able to account for the additional project value that real projects have due to the presence of management flexibility, to account for this using DCF, one would have to perform multiple simulations representing different possible management adjustments to the project. However, DCF and real options are also fundamentally different in their approach to risk discounting. Real-option valuation applies a risk-adjustment to the source of the uncertainty in the cash flow, alternatively DCF only adjusts for risk at an aggregate net cash flow level. Because of this, real options differentiates between projects based on each project's unique risk characteristics, whereas DCF does not.

Real options are based on financial options but are applied to real assets (e.g., real estate, products, intellectual property) rather than tradeable securities. Real options represent the flexibility to alter the course of action in a real assets decision, depending on future developments. Financial options represent a "side bet" that is not issued by the company whose stock is involved, but by some other entity that has no influence or connection with the company on which the bet is placed. In the case of real options, the bet is placed by the company that controls the underlying asset.

As an example of a real option, assume that company XYZ pays $20M for patent rights on a new technology. They estimate that it will cost another $100M to develop and commercialize the technology. The payoff for developing and commercializing the technology is uncertain. Buying the patent is equivalent to buying an option. Company XYZ may never invest the additional $100M, in which case the patent rights (the "option") expires. Company XYZ can wait before investing more (for the uncertainty in the payoff to reduce).

There are potentially many opportunities to apply options-based modeling to system sustainment problems, some of which are discussed in the following subsections.

10.4.1 *Technical data package procurement*

TDPs are described in Section 2.5. TDPs are expensive for customers to procure. Because of uncertainties in future needs, it is difficult to determine an appropriate price for the TDP and often customers trim budgets by not initially purchasing the TDP from contractors. If/when it becomes evident in the future that the TDP should have been acquired to facilitate the sustainment of the system, the expense of the TDP may be prohibitive and/or its access limited.

Real options allow a project to acquire the right, but not the obligation to make a transaction at or before some future point in time. In the TDP case, this could mean paying for an option that guarantees various types of future access to the TDP if desired at some a predetermined price (exercise price). When formulating TDP access via a real option the challenge becomes valuating the option to determine: the right price for the option (e.g., subscription fee), and what the exercise price(s) of the option should be, while taking into account all uncertainties in how the future could play out and the modification of risk in the project based on management decisions made. In addition, a real options model could determine when (based on need and remaining support life) to optimally exercise such an option. An options-based approach helps customers avoid costly acquisition of TDPs that may never be used, while ensuring (and appropriately pricing) the access to the data when and if necessary.

Preliminary work on the application of real options to TDP acquisition has been done in [10.8, 10.9]. These works do a good job of reviewing the state of TDP acquisition and challenges that they present. References [10.8, 10.9] formulate a renewable option to a simple example problem characterized by the need for TDP access to manage DMSMS problems.

10.4.2 *Design refresh planning*

Design refresh planning (DRP) is discussed in Section 7.5.1.3. The DRP problem requires sustainers to decide on the best time(s) and content for design refreshes. Where "best" usually means lowest life-cycle cost, least availability impact, or some combination of these.

The DRP problem could be formulated as a real options problem where management's flexibility at future points in time is to either refresh the design or continue with reactive mitigation approaches (Section 7.5.1.1), or deciding to bridge buy versus delaying refreshes (i.e., recasting the Porter design refresh costing model as a real options model).

Josias [10.10], uses a real options model to assess obsolescence management associated with strategic decisions about when to invest, what technology to invest in, waiting until a future point in time when a new technology may be available.

10.4.3 *Predictive maintenance options*

When predictive maintenance is analyzed, the operative question is often when to perform maintenance in response to a predicted RUL. The longer the predicted RUL, the more flexibility the sustainer has to manage the system, but RULs are uncertain and the longer one waits after an RUL indication, the higher the risk of the system failing before the appropriate maintenance resources are available. One method of optimizing the action to take (and when to take it) based on an uncertain RUL is using a maintenance option.

A maintenance option is a real option is defined by [10.11] as,

- Buying the option = paying to add PHM to the system (including the infrastructure to support it)
- Exercising the option = performing predictive maintenance prior to system failure after an RUL indication
- Exercise price = predictive maintenance cost
- Letting the option expire = do nothing and run the system to failure then perform corrective maintenance

The value from exercising the option is the cost avoidance (corrective versus predictive maintenance) tempered with the potential loss of unused life in system components that were removed prior to failure or the predictive maintenance revenue loss. The predictive maintenance revenue loss is relevant to systems where uptime is correlated to revenue received (e.g., energy generation systems) and is the difference between the cumulative revenue that could be earned by waiting until the end of the RUL to do maintenance versus performing the predictive maintenance at some point that is earlier than the end of the RUL. In summary, the loss that appears in the value calculation is the portion of the system's RUL that is thrown away when predictive maintenance is done prior to the end of the RUL. See [10.11] and [10.12] for the analysis of systems with maintenance options.

10.5   Final Comments on Sustainment

Under the best of circumstances, sustainment provides a framework
for assuring the financial, security, and mission-success of an enterprise (where the enterprise could be a population, company, region
or nation). However, today, sustainment is usually only recognized as
an organizational goal after it has already impacted the bottom-line
and/or the mission success of the organization, which, in many cases,
is too late. Given that society is becoming increasingly dependent on
critical systems, which impact virtually every aspect of life, the sustainment culture needs to change to make it a part of the system's
design and planning. General suggestions include [10.13]:

1. Design systems for sustainability from the beginning of the system's development.
2. Developing sustainment requirements and metrics is as critical to
   a program's success as identifying requirements for cost, schedule, and performance; but, often does not receive the requisite
   attention.
3. Socialize the concept of sustainment. Generally, universities are
   good at preparing students to design new things, but the majority
   of students receive minimal exposure to the challenges of keeping
   systems going, or the role that government policies play, in regulating sustainment.

   • We need to educate students (engineers, public policy, and
     business) to contribute to the sustainment workforce.
   • We need educate everyone – even the students that will not
     enter the sustainment workforce need to understand sustain-
     ment because all of them will become customers or stake-
     holders at some level (tax payers, policy influencers, decision
     makers, etc.). The public has to be willing to resource the
     sustainment of critical systems.

4. Leverage sustainment to create more resilient systems – resilience
   is more than just reliable hardware and fault tolerant software.
   Resilience is the intrinsic ability of a system to resist disruptions, i.e., it is the ability to provide required capability in
   the face of adversity, including adversity from non-technological
   aging and governance issues. Resilient design seeks to manage
   the uncertainties that constrain current design practices. From

an engineered systems point of view, system resilience requires all of the following:

- reliable hardware and fault tolerant software;
- resilient supply chain and workforce;
- resilient legislation or governance (rules, laws, policy);
- a resilient contract structure;
- and a resilient business model.

5. Sustainment isn't only an engineering problem. Engineering, public policy and business must all work together in order to appropriately balance risk aversion with innovation and system evolution.

The world is full of critical systems (communications, transportation, energy delivery, financial management, defense, etc.). Because these systems are expensive to replace, they often become "legacy" systems. At some point the amount of money and resources being spent on sustaining the legacy system hinders the ability invest in new systems, creating a vicious cycle in which old systems do not get replaced until they become completely unsustainable or result in a catastrophic outcome.

References

[10.1] Wohlers, T. and Gornet, T. (2014). History of additive manufacturing, Wohlers Report. 2014. Available at: http://www.wohlersassociates.com/history2014.pdf.

[10.2] Mendez, S. (2018). Marines Use 3-D Printer to Make Replacement Part for F-35 Fighter. April 19, 2018. Accessed on August 12, 2021. Available at https://www.defense.gov/Explore/News/Article/Article/1498121/.

[10.3] Paoletti, I. (2017). Mass customization with additive manufacturing: New perspectives for multi performative building components in architecture, *Procedia Engineering*. Elsevier, 23 May 2017, Available at: https://www.sciencedirect.com/science/article/pii/S1877705817317824.

[10.4] Northrop Grumman (2017). Test Marks First of Its Kind for 3-D Printed Motor Components. November 1, 2017. Accessed on August 12, 2021. Available at: https://news.northropgrumman.com/news/releases/orbital-atk-demonstrates-tactical-solid-rocket-motor-manufactured-through-additive-manufacturing.

[10.5] GE, (2021). GE receives Air Force airworthiness qualification in under a year for first metal 3D-printed, critical jet engine part, June 8, 2021. Accessed on May 18, 2022. Available at: https://www.ge.com/additive/blog/ge-receives-air-force-airworthiness-qualification-under-year-first-metal-3d-printed-critical

[10.6] Jasiulewicz-Kaczmarek, M. (2013). Sustainability: Orientation in maintenance management – Theoretical background, in *Eco-Production and Logistics. Emerging Trends and Business Practices*, Golinska P. *et al.* (eds.) (Springer–Verlag Berlin, Heidelberg), pp. 117–134.

[10.7] Myers, S. C. (1977). Determinants of corporate borrowing, *Journal of Financial Economics*, 5(2), pp. 147–175.

[10.8] McGrath, M. and Prather, C. (2016). Acquiring technical data with renewable real options, *Proceedings of the 13th Annual Acquisition Research Symposium*, April.

[10.9] Thompson, G. E. and McGrath, M. (2019). Technical data as a service (TDaaS) and the valuation of data options, *Proceedings of the 16th Annual Acquisition Research Symposium*.

[10.10] Josias, C. L. (2009). *Hedging Future Uncertainty: A Framework for Obsolescence Prediction, Proactive Mitigation and Management*, Ph.D. Dissertation, University of Massachusetts, Amherst.

[10.11] Haddad, G., Sandborn, P. A., and Pecht, M. G. (2014). Using maintenance options to maximize the benefits of prognostics for wind farms, *Wind Energy*, 17, pp. 775–791.

[10.12] Lei, X. and Sandborn, P. A. (2016). PHM-based wind turbine maintenance optimization using real options, *International Journal of Prognostics and Health Management*, 7, pp. 1–14.

[10.13] Sandborn, P. and Lucyshyn, W. (2019). Defining sustainment for engineered systems – A technology and systems view, *ASME Journal of Manufacturing Science and Engineering*, 141(2).

Appendix A

Discounted Cash Flow (DCF) Analysis

Discounted cash flow (DCF) analysis is a method of valuing a cash flow using the time value of money. A cash flow is a record of all the inflows and outflows of money over time associated with a project, process, company or asset. DCF analysis provides a way to compare different cash flows (composed of different amounts at different times). In DCF, cash flows are "discounted" depending on when in time they occur.

DCF allows a cash flow like the one in Figure A.1 to be analyzed and compared to alternative cash flows. For real situations, the cash flow may be very complex with different amounts of money coming and going at different times. In order to facilitate the apples-to-oranges comparison of different alternative cash flows one must determine the equivalent value of the cash flow at a selected point in time (most often at time zero, i.e., the "present value"), but any point in time will due.

The simple cash flow shown in Figure A.1 might, for example, represent the cash flow associated with a contract to sustain a system. The downward arrow on the left end of the timeline could be the value of a lump-sum contract received by a company to perform the sustainment, and the smaller upward arrows are the annual sustainment costs paid by the company to perform the required work. If the annual sustainment costs are $100,000/year, and the sustainment lasts for 20 years, how much money should the company collect at time zero to cover all of this? If the cost of money was zero, this is easy,

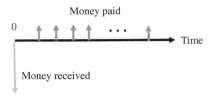

Fig. A.1.   Simple cash flow.

$$P = (20)(\$100,000) = \$2,000,000 \tag{A.1}$$

What if the cost of money is not zero? Suppose the company could invest the money it receives at time zero in the stock market and earn $r = 6\%$/year. Now how much should the company collect in order to cover the 20 years of sustainment?

$$P = \sum_{i=1}^{20} \frac{\$100,000}{(1+r)^i} = \$1,146,992 \tag{A.2}$$

The answer is less, why? If the company collects \$1,146,992 and puts it in the investment, the capital plus the interest that it earns will exactly pay for the annual sustainment costs for 20 years (i.e., the account will be drawn to zero at the end of year 20). \$1,146,992 is called the present value of the cash flow. An alternative interpretation is that I know the annual costs of sustainment, and I want to compare this to contracting the sustainment out to someone else. If the contractor is going to charge me more than \$1,146,992 (present value), performing the sustainment myself is less expensive.

A.1   Engineering Economics

Engineering economics is the application of economic principles to engineering problems, for example in comparing the costs of alternative capital projects or in determining the optimum engineering solution based on its cost. The central tool of engineering economics is DCF – it provides a structured way of comparing different cash flows. Texts on engineering economics (more generally DCF) derive various closed-form relations that can be employed to compare cash flows, e.g., Table A.1.

Table A.1. Relations for the present value of discrete cash flows.

Cash Flow Type	Present Value	Discounting Factor	Cash Flow Diagram
Single Amount	$P = F(P/F, r, n)$	$(P/F, r, n) = \dfrac{1}{(1+r)^n}$	
Uniform Series	$P = A(P/A, r, n)$	$(P/A, r, n) = \dfrac{(1+r)^n - 1}{r(1+r)^n}$	
Arithmetic Gradient	$P = G(P/G, r, n)$	$(P/G, r, n) = \dfrac{(1+r)^n - rn - 1}{r^2(1+r)^n}$	

Table A.1. (*Continued*)

Cash Flow Type	Present Value	Discounting Factor	Cash Flow Diagram

Geometric Gradient

$$P = \begin{cases} \dfrac{A_1\left[1-\left(\dfrac{1+g}{1+r}\right)^n\right]}{r-g}, & g \neq r \\[2ex] A_1\left[\dfrac{n}{1+r}\right], & g = r \end{cases}$$

Note:

P = present value

F = future value

r = discount rate or weighted average cost of capital (WACC) per time period – see Section A.2

n = number of time periods

A = annual (per period) value

G = arithmetic gradient (per period)

g = geometric gradient (per period)

A_1 = base cash flow in period 1.

All of the relations in Table A.1 assume end-of-period compounding and discrete discounting.

Engineering economics texts provide more comprehensive versions of Table A.1 that include the relations for the future values (F) and equivalent annual amounts (A). For example, using the factors in Table A.1, the solution to the simple example problem in Equation (A.2) is,

$$P = \sum_{i=1}^{20} \frac{\$100{,}000}{(1+r)^i} = P(P/A, r, n)$$

$$= \frac{(1+r)^{20} - 1}{r(1+r)^{20}} = \$1{,}146{,}992 \tag{A.3}$$

Engineering economics extends the analysis of DCFs with the inclusion of depreciation, taxation, etc. These are all important modifiers, but outside of the scope of what we need in this book. Many appropriate engineering economics texts exit, e.g., [A.1, A.2].

A.2 The Weighted Average Cost of Capital (WACC)

The inclusion of cost of money within cash flow analyses in engineering economics and life-cycle costing is a very important (and in some cases dominate) contributing factor in understanding the respective costs. Cost of money reflects the fact that the use of money to support a system (e.g., to fund design, manufacturing, and sustainment) is not free, i.e., the money has to come from some source and it is likely that that source will require some form of compensation over time for the use of their money. The cost of money is represented by r in Equations (A.2) and (A.3).

In general, there are three sources of funding available to a company to fund its operations: retained earnings (e.g., money in a bank account), borrowed money (debt financing), and the sale of equity (e.g., stocks). If the money to support a project is obtained via a loan (debt financing), then the cost of that money is the interest paid to the loan provider. If all of the money is obtained via a loan then the interest rate on the loan is set when the money is obtained and the interest rate can simply be used to modify future cash flows as in Equation (A.2), this is however, rarely the case. Usually companies are funded by, and fund projects via, a combination of debt and equity capital.

Most engineering economics texts refer to the rate paid for money as simply the "interest rate" and many engineers more generally call it the "discount rate". However, both of these terms infer the source of the money – interest rate infers debt financing, while the discount rate is defined as the interest rate charged on loans made by the Federal Reserve Bank's discount window to commercial banks and other depository institutions.

Several methods can be used to determine the cost of money, but in many cases, these methods are more art than science. A common strategy is to calculate a weighted average cost of capital (WACC). The WACC represents a weighted blending of the cost of equity and the after-tax cost of debt.

A.2.1 *The cost of equity*

Equity is a stock or any other security representing an ownership interest in a company. Companies, whether public or private, raise money by selling equity. Unlike debt, for which the company pays a set interest rate, equity does not have a predefined price. However, this doesn't mean that equity has no cost to a company. Equity holders (e.g., shareholders) expect a return on their investment in a company.

The equity holders' required rate of return represents a cost to the company, because if the company cannot provide the expected return, the equity holders may sell their equity, which will cause the stock price to drop.

The effective cost of equity is the company's cost of maintaining a share price that meets the expectations of the investors. A common method for calculating the cost of equity uses the capital asset pricing model (CAPM),[1]

$$R_e = R_f + \beta(R_m - R_f) \tag{A.4}$$

where

R_e is the cost of equity,

---

[1]Developed by William Sharpe from Stanford University who shared the 1990 Nobel Prize in Economics for the development of CAPM [A.3].

R_f is the risk-free interest rate, the interest rate of U.S. Treasury bills or the long-term bond rate is frequently used as a proxy for the risk-free rate,

R_m is the market return,

β is the sensitivity (also called volatility), and

$(R_m - R_f)$ is the R_p, equity market risk premium (EMRP).

In Equation (A.4), the sensitivity (β) models the correlation of the company's share price with the market.[2] $\beta = 1$ indicates that the company is correlated to the market; $\beta > 1$ means that the share price exaggerates the market's movements; and $\beta < 1$ means that the share price is more stable than the market. A $\beta < 0$ indicates a negative correlation with the broader market. The EMRP is the return that investors expect above the risk-free interest rate. The EMRP is the compensation that investors require for taking extra risk (above the risk-free rate) by investing in the company's stock, i.e., EMRP is the difference between the risk-free rate and the market rate. There are several commercial services that calculate the EMRP and β for public companies.

Adjustments are commonly made to the cost of equity calculated in Equation (A.4) to account for various company-specific risk factors including: the company's size, lawsuits that may be pending against the company (or lawsuits that the company has pending against others), the company's dependence on key employees, and customer base concentration. The magnitude of these adjustments are often based on investor judgment and will vary significantly from company to company.

A.2.2   *The cost of debt*

Debt is an amount of money borrowed by one party from another. Corporations use debt as a method for making large purchases that they could not afford under normal circumstances. A debt arrangement gives the borrowing party permission to borrow money under the condition that it is to be paid back at a later date, usually with interest. Compared to the cost of equity, the cost of debt is more

---

[2]Beta is the covariance between the return of the security and the return of the market divided by the variance of the market return.

straightforward to calculate. The cost of debt (R_d) is the market rate the company is paying on its debt.

Because companies benefit from tax-deductible interest payments on debt, the net cost of debt is the interest paid less the taxes paid – this is the "tax shield" that arises from the interest expense. As a result, the after-tax cost of debt is R_d (1 – corporate tax rate).

A.2.3   *Calculating the WACC*

Combining the cost of debt and equity together based on the proportion of each, we obtain the overall cost of money to the company. *WACC*, the weighted average of the cost of capital is given by

$$WACC = R_e(E/V) + R_d(1 - T_e)(D/V) \qquad (A.5)$$

where

V is the company's total value (equity + debt),
D/V is the proportion of debt (leverage ratio),
E/V is the proportion of equity, and
T_e is the effective marginal corporate tax rate.[3]

Figure A.2 shows the variation of *WACC* with the ratio of D to E. Note, in Figure A.2 that the costs of equity and debt vary

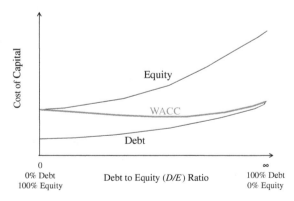

Fig. A.2.   Variation in WACC with D/E ratio.

---

[3]The *effective tax rate* is the actual taxes paid divided by earnings before taxes.

with the company's debt to equity mix, for example, the cost of debt for a company increases as more of the company is financed via debt (because lenders infer more risk and therefore charge a higher interest rate). Also note that as the cost of debt increases, the cost of equity also increases – why? The costs of debt and equity track each other because equity holders are always taking more risk than debt holders and therefore require a premium return above that of debt holders. It is also important to point out that there is an implicit assumption in Figure A.2 that the company's value does not change with the D/E ratio, which may not be the case.

In the calculation of the WACC one can subdivide the cost of equity into different types of equity, e.g., common and preferred stock. Also, sometimes the rate of return on retained earnings is also included as a separate term in Equation (A.4).

Be careful: Equation (A.5) appears easier to calculate than it actually is. No two people will calculate the same value of WACC for a company due to their unique judgments about the circumstances of the company and the valuation methods that they use.

As a simple example of computing the WACC, consider a semi-conductor manufacturer that has a capital structure that consists of 40% debt and 60% equity, with a tax rate of 30%. The borrowing rate (R_d) on the company's debt is 5%. The risk-free rate (R_f) is 2%, the β is 1.3 and the risk premium (R_p) is 8%. Using these parameters, the following can be computed:

$$R_e = R_f + \beta(R_m - R_f) = 0.02 + 1.3(0.08) = 0.124$$

$$D/V = 0.4/(0.6 + 0.4) = 0.4$$

$$E/V = 0.6/(0.6 + 0.4) = 0.6$$

$$WACC = R_e(E/V) + R_d(1 - \text{corporate tax rate})(D/V)$$

$$= 0.124(0.6) + 0.05(1 - 0.3)(0.4) = 0.0884$$

The WACC comes to 8.84% (this is a "beta-adjusted discount rate" or "risk-adjusted discount rate").

Actual values of WACC for companies vary widely. It is not uncommon for WACCs to range from 3–4% up to 20% or more.[4]

---

[4]Does the U.S. Government have a WACC? Yes, it's the rate on 3, 5, 7, 10, and longer-term treasury securities (T-Bills).

A.2.4 *Comments on WACC*

One of the biggest problems with WACC is that while it may accurately reflect what a company believes its cost of money is at the current time, the dynamics of the broader economy and the company's capital structure change with time. Therefore, the WACC is not constant over time. Specifically, the WACC is dynamic because: (1) a company's debt to equity ratio changes over time;[5] (2) the cost of equity (R_e) may change with time; (3) the cost of debt (R_d) may change over time; and (4) the tax rate (T_e) will be a function of profitability and tax breaks allowed for certain industries in certain locations during certain periods of time. Computing the WACC for a future time is difficult, but really important. Assuming that today's WACC will remain constant into the future may be a source of significant errors in life-cycle cost modeling. For example, at a macro-level, world economics dictate whether interest rates on debt rise or fall, high-profile corporate disasters increase the perceived risk of equity investments, and as technologies mature the risk that investors take in technology-oriented companies may decrease.

What engineers often call "discount rate" would be referred to as "WACC" by business analytics people. The WACC is not the inflation rate! In actuality, "WACC is neither a cost nor a required return, it is a weighted average of a cost and required return" [A.4].

A.3 Inflation

Inflation is an increase in the units of money (i.e., currency) required to purchase the same amounts of goods or services. Alternatively stated, it is the decrease in the purchasing power of money (currency), i.e., one unit of money buys less goods or services in the future.[6] The actual value (what it costs you in currency when you

---

[5]Depending on the form that the debt takes the D/E ratio may or may not remain constant. For example, the D/E ratio remains unchanged for debt in the form of a bond for which only the interest (coupon) payments are made, which is replaced by an equivalent bond at its maturity date. In the case of a loan whose balance reduces as payments are made, the D/E ratio drops over time.

[6]From the early 1990s through 2020, the inflation rate in the United States remained relatively low ($\leq 3\%$/year). In 2021 the annual inflation rate started to

actually have to pay for an item n time periods from now) is related to the real value (the equivalent value in today's currency) by

$$V_{real} = \frac{V_{actual}}{(1+f)^n} \tag{A.6}$$

where f is the inflation rate (fractional increase or decrease per time period). Although Equation (A.6) looks like Equation (A.2), it is not. The inflation rate (f) is not the same as the WACC (or discount rate, r). Equation (A.2) (and the equations in Table A.1) discount or compound value over time, but assume that there is no change in the value of the currency. In Equation (A.6), V_{actual} is the value in currency n time periods in the future and V_{real} is the equivalent value in today's currency, note in this case V_{actual} and V_{real} do not contain any discounting or compounding. The inflation rate is related to the real and market discount rates in the following way,

$$r_f = r + f + rf \tag{A.7}$$

where

  r_f is the market (or nominal) discount rate,
  r is the real discount rate, and
  f is the inflation rate.

The market rate is the rate that would be quoted to you as the interest rate paid by a bank account or the interest rate on a car loan or mortgage. The real rate is what you are really paying or really receiving after the effects of inflation have been accounted for. If the product rf is small, Equation (A.7) reduces to,

$$r \cong r_f - f \tag{A.8}$$

which is intuitively what you would expect.[7] For example, if you can obtain an interest rate of $r_f = 5\%$/year on a bank account, but the inflation rate is $f = 2\%$/year, the real rate is approximately $r = 3\%$/year (actually it is $r = 2.94\%$).

---

rise significantly as the world emerged from the COVID-19 pandemic, creating significant concern for critical system integrators delivering on firm fixed-price contracts that can't be adjusted or renegotiated.

[7]Equation (A.7) differs from the intuitive adjustment because interest earned or paid is not exempt from inflation.

There are several ways that inflation can be accounted for when analyzing cash flows. One method is to use real cash flows and discount them using the real discount rate (r). The real cash flow is composed of the real amount of money (currency) paid or received when it is paid or received (V_{real}), where the real cash flow values are found using Equation (A.6). Alternatively, an actual (nominal) cash flow (V_{actual}) can be used and then discounted using the market discount rate. The present or future value determined is same for both methods, but care must be taken not to mix nominal cash flows with real discount rates and vice versa.

References

[A.1] Blank, L. and Tarquin, A. (2018). *Engineering Economy*, 8th Ed. (McGraw Hill, USA).
[A.2] Newnan, D. G., Eschenbach, T. G., and Lavelle, J. P. (2012). *Engineering Economic Analysis*, 8th Ed. (Oxford University Press, USA).
[A.3] Sharpe, W. F. (1964). Capital asset prices – A theory of market equilibrium under conditions of risk, *Journal of Finance*, XIX(3), pp. 425–442.
[A.4] Fernandez, P. (2011). WACC: Definition, misconceptions and errors, *IESE Business School, University of Navarra, Working Paper WP-914.*

Problems

A.1 What is the present value of a single $20M expenditure 9 years from today if $r = 3.5\%$/year compounded annually?

A.2 If I invest $19M today at $r = 7\%$/year compounded annually, what will I have at the end of 8 years?

A.3 In Problem A.2, how long does it take to double your original investment?

A.4 The following two options are provided for sustaining a system for the next 20 years: Option A involves paying an annual maintenance cost of $500K/year every year, and Option B requires paying an annual maintenance cost of $250K/year every year and a single design refresh that costs $5M at the end of year 8. Which has a lower present value? Assume $r = 6\%$/year compounded annually and an end-of-year convention.

A.5 Suppose that Problem A.4 has a third option: Option C pays a third party to maintain the system and they charge $160K in year 1, which is increased by 13% each year thereafter.

A.6 Show that the summation in Equation (C.3) is $20,021.47 when $r = 0.08$/year and $\lambda = 2$ failures/year using the uniform series relation in Table A.1. Hint, you will need to find an effective r that corresponds to half a year.

A.7 Your company uses only debt and equity to finance its activities. The company has 70% debt and 30% equity. The before tax cost of debt is 9.5%/year and the tax rate is 20%/yr. The risk-free rate is 1%/year and market risk premium is 7%/year. Assume a β of 1.5. What is your company's WACC?

A.8 You are a business analyst tasked with determining the WACC (weighted average cost of capital) of a company with the following data:

$$E/V = 0.4$$
$$D/V = 0.6$$
$$\text{Cost of debt}\,(R_d) = 0.054$$
$$\text{Effective tax rate}\,(T_e) = 0.125$$
$$R_m = 0.13$$
$$R_f = 0.016.$$

The company has been public for 2 years during which time the stock and the market have a covariance of 0.03 and the variance of the market is 0.02 (hint, see footnote 2). What is the WACC of for the company?

A.9 Assuming that the inflation rate is 2%/year, how much currency will you have to pay at the end of year 8 for the design refresh in Problem A.4? (a) assuming that the $5M is an actual amount. (b) assuming that the $5M is a real amount.

A.10 Solve Problem A.4 assuming an $f = 2\%$/year inflation rate. Assume that the cash flow amounts given in Problem A.4 are real and that the discount rate given in Problem A.4 is the market rate. Stating that the cash flow in Problem A.4 is real is equivalent to stating that the cash flow values in the problem correspond to year 0 and expected to increase with inflation.

Appendix B

Monte Carlo Analysis

Uncertainty is defined as the state of having limited knowledge. Limited knowledge makes it impossible to exactly describe the existing state or future outcomes based on that state. Accounting for uncertainties is very important in all types of modeling. Models rarely predict exact answers. If your boss asks you to predict the sustainment cost of a new system during its design process and your answer is $2,345,321.45 per system per year, there is one thing that your boss knows with a 100% certainty, which is that you are wrong. Chances are that prior to the actual fielding of the system, there will be changes that will impact the system's sustainment cost, and not every instance of the system will see the same environmental stresses, fail in the same way at the same time, or cost the same amount to repair. After a population of the system has been fielded for a period of time, the sustainment cost per system is probably best represented by a probability distribution (PDF).

From a modeling standpoint, the sources of error (uncertainty) in the values predicted by models include the following[1]:

- The description of the system may not be fully known – that is, the data going into the models may be unavailable or inaccurate (data or parameter uncertainty).

---

[1]Other taxonomies and types of uncertainty, in addition to those mentioned here, may be relevant depending on the activities being considered, including measurement uncertainties and subjective uncertainties.

- The knowledge of the environment in which the system will operate may be incomplete; boundary conditions may be inaccurate or poorly understood, operational requirements may not be clear.
- The formulation of the model may be inaccurate, the understanding of the behavior of the system may be incomplete, or the model may represent a simplification of a real-world process (model uncertainty).
- Computational inaccuracies or approximations may occur. Even if the formulation of the model is accurate, numerical fitting techniques may be necessary to execute the model and the solution may only represent an approximation to the actual solution.

The uncertainty in a model can be represented as shown in Figure B.1. *Epistemic* is defined as, relating to, or involving knowledge. Epistemic uncertainties are due to a lack of knowledge. Collecting more data or knowledge can shrink epistemic uncertainties. For example, the time it takes to perform a manufacturing activity in part is an epistemic uncertainty that can be decreased if additional data collection and process observation can establish the duration of the activity, thus increasing the body of knowledge.

Aleatory (or aleatoric) means "pertaining to luck", and derives from the Latin word *alea*, referring to throwing dice. Aleatory uncertainties cannot be reduced through further observation, data collection or experimentation. Aleatory uncertainties have an inherently random nature attributable to the true heterogeneity or diversity in

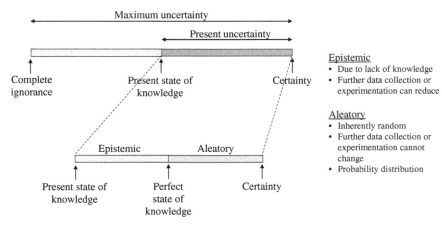

Fig. B.1. Representation of various types of uncertainty [B.1].

a population or an exposure parameter (aleatory uncertainty is also known as variability). An example of an aleatory uncertainty in a maintenance activity could be the time required for the appropriate test equipment to become available after the activity has been completed.

It is often just as important to understand the size and nature of errors in a predicted value as it is to obtain the prediction. When proposals are made, business cases constructed, and quotations prepared for sustaining systems, management needs to understand the uncertainties that are present in the prediction. Without a statement of uncertainties, a prediction is incomplete and potentially useless.

B.1 Uncertainty Modeling

Methods for sensitivity analysis and uncertainty propagation can be classified into the following four categories [B.2]: (a) sensitivity testing, (b) analytical methods, (c) sampling-based methods, and (d) computer algebra-based methods.

Sensitivity testing involves studying a model's response to a set of changes in model formulation, and for selected model parameter combinations. In this approach, the model is run for a set of sample points for the parameters of concern or with straightforward changes in model structure (e.g., in model resolution). This approach is often used to evaluate the robustness of the model, by testing whether the model response changes significantly in relation to changes in model parameters and the structural formulation of the model. The application of this approach is straightforward, and it has been widely employed. Its primary advantage is that it accommodates both qualitative and quantitative information regarding variation in the model. However, its main disadvantage is that detailed information about the uncertainties is difficult to obtain. Further, the sensitivity information depends to a great extent on the choice of the sample points, especially when only a small number of simulations can be performed.

Analytical methods involve either differentiating the model equations and subsequently solving of a set of auxiliary sensitivity equations, or reformulating the original model using stochastic algebraic/differential equations. Some of the widely used analytical methods for sensitivity/uncertainty are: (a) differential analysis methods, (b) Green's function method, (c) the spectral-based stochastic finite element method, and (d) coupled and decoupled direct methods. The analytical methods require the original model equations and may require that additional computer code be written for the solution of the auxiliary sensitivity equations – this often proves to be impractical or impossible.

Sampling-based methods involve running a set of models at a set of sample points, and establishing a relationship between inputs and outputs using the model results at the sample points. Widely used sampling-based sensitivity/uncertainty analysis methods are include: (a) Monte Carlo sampling methods (the remainder of this chapter focuses on these methods), (b) the Fourier Amplitude Sensitivity Test (FAST), (c) reliability-based methods, and (d) response-surface methods.

Computer algebra-based methods involve the direct manipulation of the computer code, typically available in the form of a high-level language code, and estimation of the sensitivity and uncertainty of model outputs with respect to model inputs. These methods do not require information about the model structure or the model equations, and use mechanical, pattern-matching algorithms to generate a "derivative code" based on the model code. One of the main computer algebra-based methods is automatic (or automated) differentiation.

B.2 Representing the Uncertainty in Parameters

In sustainment modeling, nearly every parameter that appears in the models has both an epistemic and aleatory component. As an example, consider the repair time for a particular type of system failure. Observation and data collection for 1000 repairs results in 1000 different repair times. When the repair times are plotted as a histogram, Figure B.2 is obtained.

For example, Figure B.2 indicates that if 1000 repairs are done, 0.369 or 36.9% of the repairs will have a repair time between 55 and 65 hours.

The histogram of measured results shown Figure B.2 can be fit with a known distribution type, for example this distribution can be represented as a normal distribution with a mean of 67 hours and a standard deviation of 10 hours.

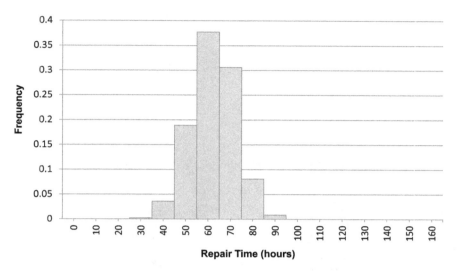

Fig. B.2. Example histogram of measured repair times.

B.3   Monte Carlo Analysis

Monte Carlo refers to a class of algorithms that rely on repeated sampling of PDFs representing input parameters to develop a histogram of results.

B.3.1   *How does Monte Carlo work?*

Suppose we have the following equation to solve:

$$G = B + C \qquad\qquad (B.1)$$

If we know the values of B and C (say $B = 2$ and $C = 3$) then G is easy to solve for. But what if we don't know exactly what B or C are – that is, there is some uncertainty associated with them. Then what is G? If we knew the range of values that B and C could take (their minimum and maximum values), we could establish the largest value and smallest value that G could have. Alternatively, the average values of B and C could be used to find the average value of G from Equation (B.1) (however, this only works if the relationship between G, B and C is linear and B and C are represented by symmetric

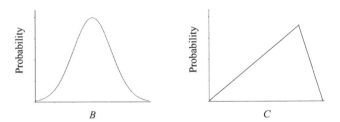

Fig. B.3.   Probability distributions representing B and C.

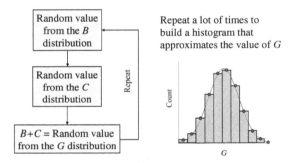

Fig. B.4.   Monte Carlo solution process.

distributions). All of these are potentially useful results, but suppose we want to know the distribution of G.

Let's generalize the problem a bit. Suppose that B and C were represented as PDFs like the ones shown in Figure B.3. It is intuitive that the resulting G (from Equation (B.1)) will also be a PDF, but how do we find it?

The Monte Carlo method of solving this problem is to sample the B and C distributions, combine the samples as prescribed in Equation (B.1) to obtain a sample of G, and then repeat the process many times to generate a histogram of G values. This process is shown in Figure B.4.

For this process to work, two key questions must be addressed. How do we sample from a distribution in a valid way? And how many times must the process in Figure B.4 be repeated in order to build a valid distribution for G?

It is worthwhile at this point to clarify some terminology. A *sample* is a collection of observed random values; one value sampled from the distribution for B and one value sampled from the distribution for C together are referred to as a single sample. Each sample can be used to independently generate one final value (one value of G). The end result of applying one sample to the Monte Carlo process is referred to as an *experiment*. The total number of samples (which corresponds to the total number of computed values of G) is referred to as the *sample size* and all the experiments together create summary statistics and a *solution*.

The Origins of Monte Carlo

Stanislaw Ulam, a mathematician who worked for John von Neumann on the Manhattan Project in the United States during World War II, is reputed to have invented the Monte Carlo method in 1946 by pondering the probabilities of winning a card game of solitaire while convalescing from an illness [B.3]. In the 1940s, scientists at Los Alamos Scientific Laboratory (today known as Los Alamos National Laboratory) were studying the distance that neutrons would travel through various materials. Analytical calculations could not be used to solve the problem because the distances depended on how the neutrons scattered during their transit through the material, an inherently random process. von Neumann and Ulam suggested that the problem be solved by modeling the system on a computer. Since the Manhattan Project was highly secret, the work required a code name. "Monte Carlo" was chosen as a reference to the Monte Carlo Casino in Monaco. Although Neumann and Ulam coined the term "Monte Carlo", such methods can be traced as far back as Buffon's needle in the 18th century.

Monte Carlo is not iterative – that is, the results of the previous experiment are not used as input to the next experiment. Each individual experiment has the same accuracy as every other experiment. The overall solution is composed of the combination of all the individual experiments. Each individual experiment in a Monte Carlo analysis can be thought of as the complete and accurate solution for one member of a large population. The end result of using many samples (each sample representing one member of the population) is a statistical representation of the population. The population could represent, for example, many instances of a system or many applications of a specific repair activity.

B.3.2 *Random sampling values from known distributions*

For Monte Carlo to work effectively, the samples obtained from the B and C distributions need to be distributed the same way that B and C are distributed. The question boils down to determining how to obtain random numbers that are distributed according to a specified distribution. For example, the *value* shown in Figure B.5 is not a uniformly distributed number, i.e., all values between 0 and 1 are not equally likely.

In order to obtain samples distributed in a specified way, we need to generate the cumulative distribution function (CDF) that corresponds to a PDF like that shown in Figure B.5. In general CDFs are found from the PDF using

$$F(x) = \int_{-\infty}^{x} f(\tau)d\tau \qquad (B.2)$$

where $f(\tau)$ is the probability density function (PDF) and x is the point at which the value of the CDF is desired, as shown in Figure B.6.

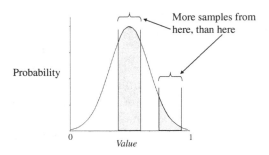

Fig. B.5.   Distributed random number.

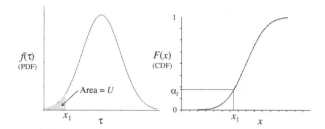

Fig. B.6. Example PDF and the corresponding CDF.

To obtain a sampled value from the distribution (the sampled value is called a *random variate* or *random deviate*), a uniformly distributed random number between 0 and 1 (inclusive) is generated. This uniform random number (U) corresponds to the fraction of the area under the PDF $(f(\tau))$ and is the value of the CDF $(F(x))$ that corresponds to the sampled value (x_1). This works because the total area under $f(\tau)$ is 1.

If a variable is represented by a PDF that has a closed-form mathematical expression for its CDF, then obtaining a sampled value from the distribution is easy. Simply choose a uniformly distributed random number between 0 and 1 inclusive and set $F(x)$ equal to it, then find the corresponding x. However, not all PDFs have closed-form CDFs. Most notably, there is no closed-form solution to Equation (B.2) for the normal distribution.[2]

The sampling strategies discussed in this chapter are referred to as transformation methods (specifically, inverse transform sampling). An alternative is called the rejection method [B.5], which does not require a CDF (it only requires that the PDF be computable up to an arbitrary scaling constant). The rejection method has the advantage of being straightforwardly applicable to multivariate PDFs. However, rejection methods are much more computationally intensive than transformation methods.

Many applications and programming languages have built in functions that can generate the CDF corresponding to various

---

[2]However, extremely efficient numerical approximations to the CDF for normal distributions do exist; see, e.g., [B.4].

distributions (even when the CDF does not have a simple closed-form expression). Table B.1 provides the CDFs and corresponding Microsoft Excel® functions for some common distributions.

B.3.3 *Random sampling from a data set*

Sometimes you have a data set that represents observations or possibly the result of an analysis that determines one of the variables in your model. You could create a histogram from the data (like Figure B.2), fit the histogram with a known distribution form, determine the CDF of the distribution (either in closed-form or numerically), and sample it as described in Section B.3.2. However, why go to the trouble of approximating a data set with a known distribution when you already have the data set? A better solution if you have a sufficiently large data set is to directly use the data set for sampling. If the data set has N data points in it,

1. Sort the date set in ascending order (smallest to largest): $(x_1, x_2, \ldots, x_N)$.
2. Choose a uniformly distributed random number between 0 and 1 inclusive (U).
3. The sampled value lies between the data point $\lfloor NU \rfloor$ and the data point $\lceil (N+1)U \rceil$.

The algorithm above works if you have a large data set, or if you have a small data set and do not have any other information. If you have just a few data points and you know what the distribution shape should be, then you are better off finding the best fit to the known distribution, then proceeding as described in Section B.3.2.

B.3.4 *Implementation challenges with Monte Carlo analysis*

There are several common issues that arise when Monte Carlo analyses are implemented.

Because of Monte Carlo's reliance on repeated use of uniformly distributed random or pseudo-random numbers, it is important that an appropriate random number generator is used. Since computers are deterministic, computer-generated numbers aren't really random.

Table B.1. Common distribution CDFs and corresponding Microsoft Excel® functions (where the RAND( ) function returns a uniformly distributed random number between 0 and 1 inclusive).

Distribution	CDF	Excel® Function to Find x
Uniform or Rectangular (a = minimum, b = maximum)	$F(x) = \begin{cases} 0 & a > x \\ \dfrac{x-a}{b-a} & a \leq x \leq b \\ 1 & x > b \end{cases}$	$= (b-a)*\text{RAND}() + a$
Normal (μ = mean, σ = standard deviation)	$F(x) = \Phi\left(\dfrac{x-\mu}{\sigma}\right)$	$= \text{NORMINV}(\text{RAND}(\ ),\ \mu, \sigma)$
Lognormal (μ = mean of $\ln(x)$, σ = standard deviation of $\ln(x)$)	$F(x) = \Phi\left(\dfrac{\ln(x)-\mu}{\sigma}\right)$	$= \text{LOGINV}(\text{RAND}(\ ),\ \mu, \sigma)$
Weibull (β = shape parameter, η = scale parameter, γ = location parameter)	$F(x) = 1 - e^{-\left(\frac{x-y}{\eta}\right)^{\beta}}$	$= (\eta*(-\text{LN}(\text{RAND}())))^{\wedge}(1/\beta) + \gamma$
Symmetric Triangular (a = maximum, b = minimum)	$F(x) = \begin{cases} 2\left(\dfrac{x-a}{b-a}\right)^2 & a \leq x \leq \dfrac{a+b}{2} \\ 1 - 2\left(\dfrac{b-x}{b-a}\right)^2 & \dfrac{a+b}{2} \leq x \leq b \end{cases}$	$F = \text{RAND}()$ if $F \geq 0.5$ then $= a + (b-a)*\text{SQRT}(F/2)$ if $F > 0.5$ then $= b - (b-a)*\text{SQRT}((1-F)/2)$

But, various mathematical operations can be performed on a provided random number seed to generate unrelated (pseudo-random) numbers. Be careful; if you use a random number generator that requires a seed provided by you, you may get an identical sequence of random numbers if you use the same seed. Thus, for multiple experiments, different random number seeds may have to be used. Many commercial applications use a random number seed from somewhere within the computer system, commonly the time on the system clock, therefore, the seed is unlikely to be the same for two different experiments.

In general, you should not use an unknown random number generator; random number generators should be checked (see [B.6]). While it is impossible to prove definitively whether a given sequence of numbers (and the generator that produced it) is random, various tests can be run. The most commonly used test of random number generators is the chi-square test;[3] however, there are other tests – for example, the Kolmogorov-Smirnov test, the serial-correlation test, two-level tests, k-distributivity, the serial test, or the spectral test. Lastly, it is generally inadvisable to use *ad hoc* methods to improve existing random number generators.

In general, you do not want to restart your random number generator for each experiment. Another common implementation mistake is to choose a single uniform random number and use it to sample the distributions associated with all the variables in the experiment. This is a grave error if all the variables are supposed to be independent. Using the same random number to sample all the distributions effectively couples all the variables together so they are no longer

---

[3]To run a chi-square test, prepare a histogram of the observed data. Count the number of observations in each "bin" (O_j for the jth bin). Then compute the following:

$$D_P = \sum_{j=1}^{k} \frac{(O_j - E_j)^2}{E_j}, \quad E_j = \frac{\sum_{j=1}^{k} O_j}{k}$$

Since we are interested in the goodness-of-fit to a distribution made up of perfectly random results, the expected frequencies (E_j for the jth bin) are the same for every bin (j) and are equal to the total number of observations divided by the number of bins. D_P asymptotically approaches a chi-square distribution with $k-1$ degrees of freedom, and if $D_P < \chi^2_{\alpha,\nu}$, then the observations are random with a 1-α confidence ($\nu = k - 1$, the degrees of freedom).

independent. Doing this effectively makes the correlation coefficient between all the variables equal to one. Independent variables need to be sampled using independent random numbers.

Some distributions can produce non-physical values – that is, the tails of the distributions matter. A prime culprit is the normal distribution. Normal distributions may be problematic for parameters that cannot take on negative values since the left tail of a normal distribution goes to $-\infty$. Normal distributions can also be problematic for parameters that cannot be greater than 1 (e.g., an availability), since the right tail goes to $+\infty$. You may think that if the mean is large enough and/or the standard deviation is small enough, unrealistic numbers won't be generated; however, a few bad samples can skew the results of the analysis. It is tempting to simply screen the samples taken from the distributions and, if they are negative (for example), simply sample again; however, this practice does not produce valid distributions. Don't do it![4] Other distributions may be preferred that have controllable minimum and/or maximum values, such as triangular distributions.

Many simple tests are possible to verify the implementation of a Monte Carlo analysis model. A histogram of the values sampled can be plotted from the input distributions to verify that the sampled values result in the same distribution as the input. If the problem is linear (like Equation (B.1)) and symmetric input distributions (e.g., for B and C) are used, then the mean value of the resulting G distribution should be equal to the G calculated using the mean values of B and C. A distribution of the mean output from each Monte Carlo solution should always be normal (if the sample size is large enough – see Section B.4).

B.4   Sample Size Estimation

A fundamental question with Monte Carlo analysis is how many samples must be produced (or experiments must be performed) to generate an acceptable solution? The sample size (n) is the quantity of data points or observations that need to be collected from a Monte

---

[4]Note that there are mathematically valid truncated normal distributions that are bounded below and/or above. For an example, see [B.7].

Carlo analysis to form a solution. Because Monte Carlo is a stochastic method, we will get a different answer every time we perform the analysis. As the sample size increases, the difference in the summary statistics between repeated solutions decreases (but will never go to zero).

There are two ways to approach answering the sample size question. The practical answer is that you need to run experiments until the quantity you want from your analysis – that is, the precision of the estimate of the mean or precision of the estimate of the cumulative distribution – stops changing (or the amount of change drops below some predetermined threshold). As long as the uniform random number generator is not reset or does not otherwise begin repeating random numbers, more experiments can be run and added to the experiments you already have. For example, when you run 100 more experiments and the summary statistics you are interested in are no longer changing as you perform more experiments, then you are done. If the summary statistics are still changing, run more experiments.

Of course, the sampling problem can also be treated in a mathematically rigorous way. The sample mean is an estimation of the mean of the true population (the true population would be generated if you ran an infinite number of experiments). So how accurate is this estimation for a non-infinite number of samples? Generally, the mean is not the same when the analysis is repeated or more samples are added.

If you repeat the Monte Carlo simulation and, in each simulation, draw n samples from a population, which can be represented as an unknown distribution, based on the Central Limit Theorem, the sample means μ will follow a normal distribution. The Central Limit Theorem states that if n random samples are selected from a population with sample mean μ and sample standard deviation σ, as the sample size n becomes large, the sample means approach a normal distribution with a population mean and a standard deviation equal to $\sigma/\sqrt{n}$ (which is referred to as the standard error of the mean). This is independent of the shape of the distributions associated with the parameters in the original Monte Carlo analysis that was used to create the population.

Therefore, the standard error is a useful indicator of how close the estimate from the Monte Carlo solution is to the unknown estimand (the parameter being estimated),

$$\text{Er}(\mu) = z \frac{\sigma}{\sqrt{n}} \qquad (B.3)$$

where z is the standard normal statistic (i.e., the distance from the sample mean to the population mean in units of standard error) tabulated in statistics books.[5]

As an example, if I model the repair cost of system using 1000 Monte Carlo experiments and the mean value (μ) is 20 hours with a standard deviation (σ) of 10 hours, how many total Monte Carlo experiments will I need to run to achieve 1% accuracy with a 95% confidence? Solving (B.3) for n gives us,

$$n = \left(z \frac{\sigma}{\text{Er}(\mu)} \right)^2 \qquad (B.4)$$

For our example, $\text{Er}(\mu)/\mu = 0.01$, the value of z that corresponds to 95% confidence (assuming that the 95% is symmetric around the mean) is given by $z = \text{NORMSINV}(0.95 + 0.5 * (1 - 0.95)) = 1.645$ and the final value of $n = 6764$. In this case, if you have already run 1000 experiments, you need to run 5764 more experiments.

B.5 Example Monte Carlo Analysis

In this section we present a simple analysis performed using the Monte Carlo method. Suppose a part in a system has a time-to-failure that is given by a Weibull distribution with $\beta = 2$, $\eta = 23$ years, and $\gamma = 5$ years. If I field a system containing the component, what will the mean time-to-failure of the component be? Assuming that every time the component fails it will be immediately replaced with a new component, the mean of a 3-parameter Weibull distribution is given by,

---

[5]In Excel, the value $z = \text{NORMSINV}(\text{confidence level})$.

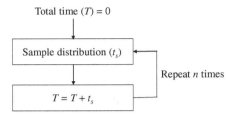

Fig. B.7. Example sampling procedure.

$$\mu = \gamma + \eta\Gamma\left(1 + \frac{1}{\beta}\right) = 25.3832\,\text{years} \tag{B.5}$$

We can also solve this with a Monte Carlo simulation. The flowchart in Figure B.7 provides the Monte Carlo simulation process we will use. In this case, the distribution is sampled by solving Equation (3.20) for t,

$$t_s = \eta[-\ln(1 - F)]^{1/\beta} + \gamma \tag{B.6}$$

Note, Equation (B.6) is also given in Table B.1. When the sampling process in Figure B.7 is complete, we can divide T by n to get the mean time to first failure for this component ($\mu = 25.5131$ years for our 10,000-experiment case). If we keep all the t_s values generated (n of them) and plot them in a histogram we get Figure B.8. The histogram in Figure B.8 is simply the distribution of failures that we started the problem with. With the Monte Carlo result for this example we can also estimate the confidence. If we specified that we want the time to first failure with a 90% confidence, we can simply sort the $n = 10,000$ samples into ascending order and count through the first 9000 of them. If we do this, we get a first failure time of 40 years, i.e., 90% of the 10,000 samples fail before 40 years.

Obviously, this is a very simple example, but it serves to illustrate the concept.

B.6 Stratified Sampling (Latin Hypercube)

The methodology considered so far in this chapter assumes random sampling from the prescribed distributions – that is, we are using

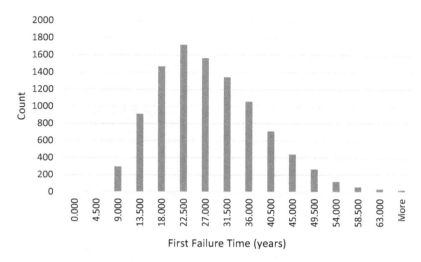

Fig. B.8.   Histogram of failure times (t_s), $n = 10,000$.

uniformly distributed random numbers between 0 and 1 inclusive to extract distributed random numbers.

Stratified sampling can characterize the population equally as well as simple random sampling, but with a smaller sample size. In stratified sampling, the data is collected to occupy prearranged categories or strata. The form of stratified sampling we are going to consider in this section is called Latin Hypercube (other stratified sampling approaches exist).

B.6.1   *Building a Latin hypercube sample (LHS)*

To building a Latin hypercube sample, four steps are required [B.8]:

1. The range of each variable is divided into n_I non-overlapping intervals each representing equal probability.
2. One value from each interval for each variable is selected using random sampling.
3. The n_I values obtained for each variable are paired in a random manner to form $n_I k$-tuplets (the LHS).
4. The LHS is used as the data to determine the overall solution.

First the range of each variable is divided into n_I non-overlapping intervals, each representing equal probability, as shown in Figure B.9.

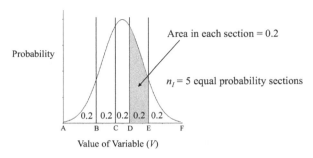

Fig. B.9. Division of the PDF into n_I equal probability intervals.

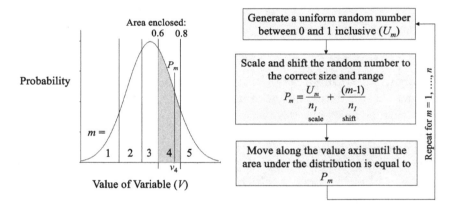

Fig. B.10. Selecting one value from each interval via random sampling.

In this example, the range of the variable V is divided into $n_I = 5$ equal probability (0.2) intervals.

Next, one value from each interval for each variable is selected using random sampling, as shown in Figure B.10. The sampling from each interval is performed essentially identically to the random sampling discussed in Section B.2.2.

In the third step, the n_I values $(v_1, \ldots, v_{n_I})$ obtained for each variable are paired in a random manner (equally likely combinations) forming $n_I k$-tuplets (k is the number of variables considered), this is called the Latin hypercube sample (LHS). For $k = 2$ (two variables, V and Z with distributions) and $n_I = 5$ intervals, we pair two random

Table B.2. Two 5-tuplets that define the LHS for a problem with two random variables (V and Z).

Computer run number	Interval used for V	Interval used for Z
1	3	2
2	1	4
3	5	1
4	2	3
5	4	5

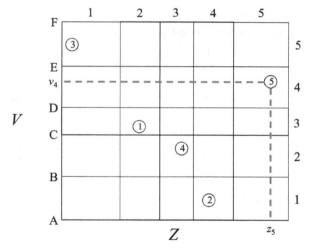

Fig. B.11. Two-dimensional representation of one possible LHS of size 5 with two variables.

permutations of $(1, 2, 3, 4, 5)$: Permutation Set 1: $(3, 1, 5, 2, 4)$ and Permutation Set 2: $(2, 4, 1, 3, 5)$, as shown in Table B.2.

Figure B.11 shows a representation of the LHS of size 5 for V and Z. Note that only the generation of the V values was shown in Figure B.9, Z is another variable with a similar generation process. In Figure B.11 v_4 is the $m = 4$ interval sample from the variable V and z_5 is the $m = 5$ interval sample from the variable Z. In general, Figure B.11 would be k dimensional and have n_I^k cells in it and produce $n_I k$-tuplets of data.

Finally, we use the LHS as the data to determine the overall solution. The data pairs specified by Table B.2 are used: (v_3, z_2), (v_1, z_4), (v_5, z_1), (v_2, z_3), (v_4, z_5). These five data pairs are used to produce five possible solutions.

B.6.2 *Comments on LHS*

LHS forms a random sample of size n_I that appropriately covers the entire probability space. LHS results in a smoother sampling of the PDFs – that is, it produces more evenly distributed (in probability) random values and reduces the occurrence of less likely combinations (e.g., combinations where all the input variables come from the tails of their respective distributions). Random sampling required n samples (n is the sample size from Section B.3) of k variables $= kn$ total samples. LHS requires n_I samples (intervals) of k variables $= kn_I$ total samples. It is not unusual for LHS to require only a fifth as many experiments as Monte Carlo with simple random sampling.

To determine n_I, apply the standard error on the mean criteria (e.g., Equation (B.3)) to each interval.

Even though variables are sampled independently and paired randomly, the sample correlation coefficient of the n_I k-tuplets of variables, in general, is not zero (due to sampling fluctuations). Restricting the way in which variables can be paired can be used to induce a user-specified correlation among selected input variables. See [B.9] for more discussion.

B.7 Discussion

Monte Carlo simulation methods are particularly useful for studying systems that have a large number of coupled degrees of freedom. Monte Carlo methods are also useful for modeling systems with highly uncertain inputs. Monte Carlo methods are not deterministic (i.e., there is no set of closed-form equations to solve for an answer).

Monte Carlo is independent of the formulation of the model – for example, the model does not have to be linear. Monte Carlo also does not constrain what form the distributions take, and the distributions need not necessarily even have a mathematical representation. Monte Carlo also has the advantage that even though it is computationally intensive, it will always work.

The main argument against Monte Carlo is that it is a "brute force" computationally intensive solution. Another potential drawback is that Monte Carlo implicitly assumes that all the parameters are independent. Correlation of the parameters in Monte Carlo analyses can be done. In general, the parameters are uncorrelated because independent random numbers are used to generate the samples. The degree to which the parameters are correlated depends on the how correlated the random numbers used to sample them are (see, e.g., [B.10]).

There are many software packages for performing Monte Carlo analysis today – Palisade, @Risk®, Minitab, and Crystal Ball® are available for Excel. A treatment of Monte Carlo implementation within Excel is provided in [B.11].

References

[B.1] Aughenbaugh, J. M. and Paredis, C. J. J. (2005). The value of using imprecise probabilities in engineering design, *Proceedings of the ASME Design Engineering Technical Conference (DETC)*.

[B.2] Isukapalli, S. S. (1999). *Uncertainty Analysis of Transport-Transformation Models*, Ph.D. Dissertation, The State University of New Jersey at Rutgers. Available at: http://www.ccl.rutgers.edu/ccl-files/theses/Isukapalli_1999.pdf. Accessed April 22, 2016.

[B.3] Eckhardt, R. (1987). Stan Ulam, John von Neumann, and the Monte Carlo method, *Los Alamos Science, Special Issue*, 15, pp. 131–137.

[B.4] West, G. (2005). Better approximations to cumulative normal functions, *Wilmott Magazine*, 9, pp. 70–76. https://lyle.smu.edu/~aleskovs/emis/sqc2/accuratecumnorm.pdf. Accessed May 8, 2016.

[B.5] von Neumann, J. (1951). Various techniques used in connection with random digits, *National Bureau of Standards Applied Mathematics Series*, 12, pp. 36–38.

[B.6] Park, S. K. and Miller, K. W. (1988). Random number generators: Good ones are hard to find, *Communications of the ACM*, 31(10), pp. 1192–1201.

[B.7] Greene, W. H. (2003). *Econometric Analysis*, 5th Ed. (Prentice Hall, USA).

[B.8] McKay, M. D., Conover, W. J., and Beckman, R. J. (1979). A comparison of three methods for selecting values of input variables in the analysis of output from a computer code, *Technometrics*, 21(2), pp. 239–245.

[B.9] Iman, R. L. and Conover, W. J. (1982). A distribution-free approach to inducing rank correlation among input variables, *Communications in Statistics*, B11(3), pp. 311–334.

[B.10] Touran, A. (1992). Monte Carlo technique with correlated random variables, *Journal of Construction Engineering and Management*, 118(2), pp. 258–272.

[B.11] O'Connor, P. and Kleyner, A. (2012). Chapter 4 – Monte Carlo simulation, *Practical Reliability Engineering*, 5th Ed. (John Wiley & Sons, UK).

Problems

B.1 Given a random variable, x, with a symmetric triangular distribution defined by $a = 2$ and $b = 4$. Write a Monte Carlo model that results in the original distribution function.

B.2 The probability of rolling seven (when rolling two die) is equal to $6/36 = 0.167$. Demonstrate this using a Monte Carlo model.

B.3 Write a Monte Carlo simulation to model a biased coin as follows. When the coin is flipped, if it lands on tails it has a 50% chance of heads or tails on the next flop. However, if the coin lands on heads then it has 75% chance of landing on heads again on the next flip. The Monte Carlo simulation should determine the probability of the coin landing on heads.

B.4 If a variable time is represented as a Weibull distribution ($\beta = 4$, $\eta = 10^5$ hours and $\gamma = 20{,}000$ hours) and the modeling program chooses the value of a random number (between 0 and 1, inclusive) equal to 0.27, what is the sample value that a Monte Carlo analysis will returned from the distribution? The Weibull distribution is described in Section 3.2.5.

Application-specific Monte Carlo problems appear throughout the other chapters of this book and in particular in Appendix C.

Appendix C

Discrete-Event Simulation (DES)

The life of a system can be abstracted into a series of specific events that occur at specific points in time. When the events occur, they change the state of the system. In some cases, the way the state of the system changes depends on the sequence of the events and/or when (in time) they occur.

For example, life-cycle cost modeling (over long periods of time) is highly dependent on when events occur because the present value of the event depends on the WACC (discount rate) and the impact of the WACC on present value depends on how far in the future the event takes place – see Appendix A. In some cases, spares may have shelf lives, so their viability depends on when the demand for them occurs. For critical systems the time-dependent quantities usually involve the operation and support of the system.

When we simulate a system that evolves over time, the system either changes continuously or discontinuously. An example of a continuous system is the weather – temperature, humidity, wind speed, etc., all change in a continuous way. Other types of systems change in a discontinuous (or discrete) manner, for example, an inventory system that decreases or increases at specific points in time when parts are demanded or replenished. In this case a graph of the quantity of parts in the inventory as a function of time would look like a series of step functions separated by periods of time where there was no change in the inventory.

Discrete-event simulation (DES) is the process of codifying the behavior of a complex system as an ordered sequence of well-defined events. DES utilizes a mathematical/logical model of a physical system that portrays state changes at precise points in simulated time called events [C.1].[1] Discrete means that successive changes are separated by finite amounts of time, and by definition, nothing relevant to the model state changes between events.

Time may be modeled in a variety of ways within the DES. Alternate treatments of time include: time divided into equal increments, i.e., a time step; unequal increments; or cyclical (periodic), e.g., as in a traffic light or bus schedule. In DES the system "clock" jumps from one event to the next, and the periods between events are ignored. A timeline is defined as a sequence of events and the times that they occur.

At each event, various properties of the system can be calculated and accumulated. Accumulated parameters of interest as the simulation proceeds along the timeline could be: time (system "clock"), cost, system up or down time, inventory levels, spares demanded, throughput, defects, resources consumed (material, energy, etc.), waste generated, etc. Using the accumulated parameters, one could generate various important results as a function of time: total cost, resources consumed, availability, readiness, spare parts consumed, return on investment, etc. Everything in the DES is uncertain and can be represented by probability distributions. This means that we model the timeline (and accumulate relevant parameters) many times (i.e., through many possible time histories) in order to build a statistical model of how the system will evolve.

C.1 Events

An event represents something that happens to the system at an instant in time that may change the state of the system where, by definition, nothing relevant to the state of the system changes

---

[1] It is difficult to pinpoint the exact origin of DES, however Conway, Johnson and Maxwell's 1959 paper [C.2] discusses many of the key points of a DES, including managing the event list (they call it an element-clock) and methods for locating the next event. It is evident that many of the concepts of DES were being practiced in industry in the late 1950s.

between events. Relevant types of events in the life cycle of a complex electronic system include:

- Scheduled maintenance (preventative)
- Unscheduled/corrective maintenance (failures followed by maintenance)
- Spares purchases (replenishments)
- Upgrades (or other scheduled system changes)
- Cost charges (e.g., inventory holding)

Events have various properties that include: costs and durations (even though events occur at an instant in time, they can have a non-zero duration), other resource requirements (workforce, facilities, administrative), resource consumption (spare parts, energy, various materials), waste generation, and possibly defects contributions (or subtractions).

An important note here is that each event is dependent of the previous events that have occurred on the timeline. The dependency may simply be timing (see the examples in Section C.2), or it may be more complex – the previous events may change the state of the system in such a way as to influence the type of event that occurs next.

Events may have start and end times if the events are not instantaneous (see Problem C.3).

C.2   Discrete-Event Simulation Examples

This section presents several DES examples beginning with a very simple example followed by more complex examples that can be used to analyze the life cycle of a system. Several cases referred to in Chapters 4 and 5 are also included.

C.2.1   *A trivial DES example*

Assume that we have some type of system whose failure rate is constant. The reliability of the system is given by Equation (3.17) as,

$$R(t) = e^{-\lambda t} \tag{C.1}$$

where t is time and λ is the failure rate. As defined in Equation (3.18) the mean time between failures for this system is $1/\lambda$ (known as the $MTBF$). Suppose, for simplicity, failures of this system are resolved instantaneous at a maintenance cost of \$1000/failure. If we wish to support the system for 20 years, how much will it cost? Assuming that the discount rate is zero, this is a trivial calculation:

$$Total\,Cost = 1000(20\lambda) \qquad\qquad (C.2)$$

The term in parentheses is the total number of failures in 20 years. If $\lambda = 2$ failures per year, the *Total Cost* is \$40,000.

This example is very easy and we certainly do not need any sort of simulation to solve it, but what if the discount rate (r) was 8%/year compounded discretely? Now the solution becomes a sum (see Appendix A), because each maintenance event has a different present value,

$$Total\,Cost = \sum_{i=1}^{20\lambda} \frac{1000}{(1+r)^{i/2}} \qquad\qquad (C.3)$$

where $i/2$ is the event date in years.[2] The *Total Cost* is now \$20,021.47 in year 0 dollars (present value).[3]

Even though the two cases described so far are pretty easy and we don't need DES to solve them, let's use DES to illustrate the process. To create a DES for these simple cases, we start at time 0 with a cumulative cost of 0, advance the simulator to the first failure event, determine the cost of that event and add it to the cumulative cost, and then repeat the process until we reach 20 years. Table C.1 shows the DES events and costs.

Obviously, there are several implicit assumptions about exactly when the events take place and other things. At this point Table C.1 is just a rather arduous way of performing the calculations in Equations (C.2) and (C.3). However, Table C.1 is a DES. In this case each failure has a specific date on which a maintenance cost is charged and added to the cumulative maintenance cost. Nothing (that costs money) is assumed to happen to the system between events.

_____

[2]The $i/2$ assumes that $\lambda = 2$ and the failures are uniformly distributed throughout the year.

[3]Problem A.6 in Appendix A asks you to solve Equation (C.3) using a uniform series cash flow from Table A.1.

Table C.1. Simple example described in terms of events.

Event Number	Event Date (years)	Event Cost $(r = 0)$	Cumulative Cost $(r = 0)$	Event Cost $(r = 8\%)$	Cumulative Cost $(r = 8\%)$
0	0	0	0	0	0
1	0.5	$1000	$1000	$962.25	$962.25
2	1	$1000	$2000	$925.93	$1888.18
3	1.5	$1000	$3000	$890.97	$2779.15
. . . .					
40	20	$1000	$40,000	$214.55	$20,021.47

C.2.2   *A less trivial DES example*

Suppose that the actual event dates in the example presented in the previous sub-section are not known, rather the time-to-failures are represented by a failure distribution. For our simple case, the corresponding failure distribution is given by Equation (3.15),

$$f(t) = \lambda e^{-\lambda t} \tag{C.4}$$

Now, instead of assuming that the failures of the system take place at exactly *MTBF* intervals (the *MTBF* is just the expectation value of the time-to-failure), they take place at intervals determined by sampling (using Monte Carlo) the $F(t)$ distribution given in Equation (3.16). Now the total cost is given by the sum in Equation (C.3), but the event dates come from sampling; so, there is no simple analytical sum to use for the solution.

Let's solve this problem using DES. First, we need to generate the failure times. For this we use the CDF of the exponential distribution from Equation (3.16),

$$F(t) = 1 - e^{-\lambda t} \tag{C.5}$$

Rearranging Equation (C.5) to solve for t we get,

$$t = \frac{-\ln[1 - F(t)]}{\lambda} \tag{C.6}$$

To sample this, we choose a random number between 0 and 1 (inclusive) that we assign to $F(t)$, then solve Equation (C.6) for t, which is the failure time sampled from the exponential distribution. Note, t

Table C.2. Time-to-failure distribution sampling example.

Event Number	Random Number $(F(t))$	Time-to-Failure Sample (years) (t)	Event Date (years) (t_c)	Event Cost $(Cost_i)$	Cumulative Cost
0	—	—	0	0	0
1	0.194981	0.108445	0.108445	$991.69	$991.69
2	0.430298	0.281321	0.389765	$970.45	$1,962.14
3	0.978275	1.914642	2.304407	$837.49	$2,799.62
...					
41	0.197316	0.109897	18.85356	$234.34	$20,826.08
42	0.971349	1.776292	20.62985	$204.40	$21,030.48

is not the next event date, it is the time measured from the previous event, so the values of t need to be accumulated to produce the event dates.[4]

Using the event dates, we can now calculate the individual event costs using,

$$Cost_i = \frac{1000}{(1+r)^{t_c}} \qquad (C.7)$$

where t_c is the cumulative failure time at the ith event. Table C.2 shows an example of the first three events and two final events in the process.[5]

In this case there is no set number of events that need to be generated to reach the 20-year support life considered in this problem; i.e., you may need more or less than 40 events to get there, so the simulation needs a stopping criteria, which is stop when the Event Date > 20 years and do not include the cost of the final event since it is after the end of support of the system. In this example, the total cost is $20,826.08.

The example described in this section samples reliability distributions to generate a sequence of events. The sequence of failure events generated in Table C.2 represents one possible future scenario

---

[4]You may not need to manually sample the distribution as we have done in Equations (C.5) and (C.6). Excel®, R, MATLAB® and other environments, for example, have commands that will directly return a sample from an exponential distribution.

[5]Table C.2 is an example. Each time we run the analysis it comes out differently because we are sampling the time-to-failure distribution.

("path") for the system. Embedding this process within a broader Monte Carlo analysis allows the generation of many future paths for the system.

C.2.3 *Corrective maintenance spares calculation*

In Section 4.4.1.2 we calculated the average number of spares needed per year for a bus example. The relevant data that associated with the bus problem is:

- 200,000 miles/year
- Constant failure rate with $\lambda = 1.4 \times 10^{-5}$ failures/mile
- Corrective maintenance (CM) takes 5 days (2740 miles of lost bus usage)
- Assume that the replacement components are "as-good-as-new"
- The failure mechanism only accumulates damage while the bus is operating (not while it is being repaired).

Let's solve with the simplest assumption first – assume that maintenance requires no time. The solution for this assumption is 2.8 renewals/year (previously calculated in Section 4.4.1.2). The DES solution follows the flow chart shown in Figure C.1. The flowchart in Figure C.1 represents the calculation of a single solution based on a single (unique) sequence of failures. The process samples for a failure, consumes a spare to restore the system, samples that component's failure again, and keeps going until the next failure is beyond 200,000 miles. The value of k at the end of this process is the number of spares required. If, instead of using Equation (C.6), the next failure time is simple set to $1/\lambda = 71,428.6$ miles, the flowchart in Figure C.1 will result in $k = 2$ (which is not the same as 2.8, that was solved for in Chapter 4). To get 2.8 we need to realize two things: (1) the starting conditions for the component are relevant, in other words not every 1-year period starts with the component good-as-new; and (2) not every year's time history is the same. If we run the DES (with the fixed failure rate) for 500 years and find the average number of spares per year, the answer will be 2.8.[6]

In reality, the failures are exponentially distributed for the bus example. This requires that we sample the time to failures, which

---

[6]To do this, change the end of support (*EOS*) in Figure C.1 from 200,000 to (200,000)(500) in order to run the simulation for 500 years. You can then take

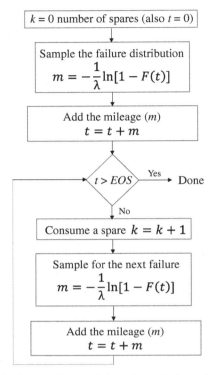

Fig. C.1.   Spares calculation flowchart (one-time history shown). *EOS* and *t* are measured in miles.

is given by Equation (C.6). The sampling is shown in Figure C.1, where $F(t)$ is a uniformly distributed random number between 0 and 1 inclusive. If the flowchart in Figure C.1 is embedded within a Monte Carlo process, i.e., run it 1000 times for 500 years, and the resulting values of k (divided by 500) are averaged, the result will be, very close to 2.8.

Now add the non-zero maintenance miles (2740) to the problem by adding 2740 to the cumulative t after every maintenance event. Again, running the flowchart 1000 times for 500 years, the value of k (per year) is close to 2.697. A histogram of the results of this simulation for 30,000 Monte Carlo experiments is shown in Figure C.2.

---

the resulting cumulative value of k and divide it by 500 to obtain the expected number of spares per year for the bus.

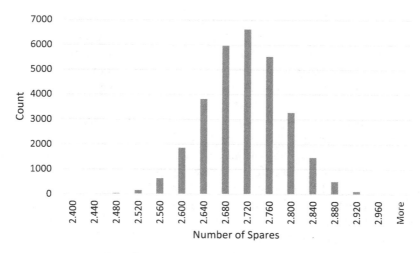

Fig. C.2.   Histogram of sparing results.

How do we determine the number of spares needed to satisfy a particular confidence level? Simply count the number of results in the data set that satisfy the confidence criteria. For our data set, 2.788 spares represents 90.36% confidence of having enough spares, i.e., 90.36% of the results generated require 2.788 or fewer spares.

C.2.4   *Corrective and preventative maintenance events calculation*

In Section 4.5.2 we calculated the average number of CM and preventative maintenance (PM) events per year for the bus example. The relevant data that associated with the bus problem is:

- 200,000 miles/year
- Weibull failure rate with $\beta = 2$, $\eta = 74{,}000$ miles and $\gamma = 0$
- CM takes 5 days (2740 miles of lost bus usage)
- PM takes 1 day (550 miles of lost bus usage)
- Assumed the PM interval is $t_p = 65{,}500$ miles
- Assume that the replacement components are "as-good-as-new"
- The relevant failure mechanism only accumulates damage while the bus is operating (not while it is being repaired).

Again, our DES needs to emulate the life cycle of this system as shown in the flow chart in Figure C.3. In this case we start by sampling the reliability distribution to find the first failure (Failure Miles). If that failure occurs before 65,500 miles, then we have a CM event; it if occurs after 65,500 miles it is a PM event at 65,500 miles. The process continues until the mileage of the next failure exceeds 200,000 miles. One key thing to point out, PM events take place 65,500 miles after the last event (a CM or PM), they do not automatically occur every 65,500 miles. Like with the example in the last section, we must account for the starting point for the bus. If we run the analysis in Figure C.3 1 year, we are implicitly assuming that the component is good-as-new at the beginning of that year. In this case we obtain 1.976 CMs and 1.498 PMs. Alternatively, if you run it for 100 years (and divide the resulting number of CMs and PMs by 100), you will get an average of 2 CMs and 1.7 PMs per year per bus.

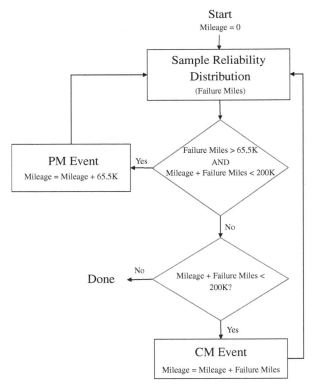

Fig. C.3. CM and PM event calculation flowchart (one time history shown).

This case effectively accounts for a randomized amount of life consumed in the component at the start of each year.

The availability, in this case, determined via the discrete-event simulation is 0.9688.

C.3   Discussion

Other approaches exist for modeling the dynamics of systems, e.g., Markov chains (see Section 5.1.7). DES is a scenario-based simulation method that simulates each item of the system separately through different event-paths/sample paths. In general, simulation-based methods consider components with differing attributes that move from one event to another in time considering component properties such as: age, maintenance history, and usage profile. Many analyses use simulation to optimize stochastic problems. Simulation-based approaches are especially useful and common when the model grows in size or the integration of multiple disciplines is required. Monte Carlo sampling is usually used for sampling from probability distributions of each parameter, as long as one can estimate reasonable distributions.

Discrete-event simulators represent a straightforward method of solving many real-world problems – they are effectively an emulation of the real world. The arguments against using DES are that they can become cumbersome and lead to long simulations for large systems – because they are "brute force" emulations of the real world, they need not oversimplify a problem in order to obtain a solution. DES can be used to find practical optimums to problems, but cannot be used to obtain provable optima.

In the example cases in Section C.2, only one type of event generating action was present (a system failure for which a maintenance action was necessary). For real systems, multiple types of events may occur concurrently on the same timeline. If one is only interested in the final state at the end of a defined period of time, then it may be possible to simulate separate independent timelines and simply accumulate the final results together. However, if one wishes to see the cost as a function of time, or if the different types of events are not independent (i.e., if the next event of type A depends on the occurrence and/or timing of an event of a type B), then the timeline has to be modeled sequentially and concurrently for all events. An example of this would be multiple system instances drawing spares from a common inventory. In this case, separate DESs for each system instance cannot be generated and then added because the timing of spare replenishment (which represents an event) depends on the demands from all of the system instances. In this case, all system instances have to be simulated concurrently.

Agent-Based Modeling (ABM)

Although DES is widely used for sustainment applications, there are alternative modeling approaches that may be advantageous under various circumstances. In particular, DES does not lend itself to modeling systems that are highly influenced by the behavior of companies or individuals.

A discrete-event simulator (DES) model assumes that the final state of a system can be determined by modeling a sequence of events in time where each event causes a change in the system state. If there are well defined models for predicting the sequence of events, and you know how each event changes the system's state, then DES works well. If, on the other hand, the final state of system is a result of the behavior of a group of independent agents, where each agent has their own set of goals and behaviors (and their own thread of control), then the aggregate system behavior can then be inferred from the state of the agents. This is called agent-based modeling (ABM).

Unlike simple objects used in DES, agents are capable of making autonomous decisions (i.e., they are able to take flexible action in reaction their environment) and agents are capable of showing proactive behavior (i.e., actions depend on motivations generated from their internal state).

Consider a simple prey/predator model example. In ABM the agents are the prey (they want to find grass to eat, not get eaten by the predator, and survive long enough to reproduce) and the predator (wants to find enough food in the form of the prey to allow it to live long enough to reproduce). At each time step,

(Continued)

> *(Continued)*
>
> predators try to move toward the prey and the prey tries to move toward the grass and away from the predators. If the predator gets close enough to the prey, the prey gets eaten. If the prey can survive long enough it reproduces. If the predator can find enough prey to eat it reproduces. Predicting the future of this system is well suited to ABM. Could you model this system with DES? Yes, but you would have to know the probability distributions associated with prey finding grass and predators finding prey, etc., in every time period. A DES model would also not be well suited to accommodating predators and prey that learn and modify their behavior over time.

Discrete-event simulators also suffer from the constraint that they only operate in one direction, i.e., forward in time. Because of this, there are many outputs of discrete-event simulators that are straight-forward to generate (e.g., cost and availability) that become very difficult to use as inputs to a design process. For example, availability requirements can be satisfied by running discrete-event simulators in the forward direction (forward in time) for many permutations of the system parameters and then selecting the inputs that generate the required availability output. Such "brute force" search-based approaches may be computationally impractical for real problems (particularly for real-time problems). There have been attempts to perform reverse simulation (run discrete-event simulators backwards in time) but this has only been demonstrated on extremely simple problems with limited applicability to the real-world systems.

References

[C.1] Nance, R. E. (1993). A History of Discrete Event Simulation Programming Languages, TR 93-21, Virginia Polytechnic Institute and State University, Department of Computer Science.

[C.2] Conway, R. W., Johnson, B. M., and Maxwell, W. L. (1959). Some problems in digital systems simulation, *Management Science*, 6(1), pp. 92–110.

Problems

C.1 In the simple example in Section C.2.1, several implicit assumptions were made about when failures occur and how they have to be fixed. Identify and discuss these assumptions.

C.2 Rework the example in Section C.2.2 assuming that the time to failure is given by a Weibull distribution with the following parameters: location parameter $= 500$ hours, shape parameter $= 4$, and the scale parameter $= 10,000$ hours.

C.3 Rework the example in Section C.2.2 (with the constant failure rate), assuming that the time to resolve the failures (which was previously assumed to be instantaneous) is given by a triangular distribution with a lower bound of 30 days, an upper bound of 60 days and a mode of 45 days. Is the cumulative cost larger or smaller than the cumulative cost when the failures are resolved instantaneously?

C.4 Calculate the final (after 20 years) time-based availability of the system in Problem C.3.

C.5 What if an infrastructure charge of $150/month is incurred in the example in Section C.2.2 (with the constant failure rate)? What is the total cost after 20 years? Hint: the infrastructure charge represents an event that is independent of the maintenance events.

C.6 Starting with the example in Section C.2.2 (with the constant failure rate), assume that each maintenance event requires one spare. For simplicity, assume that the spare costs $1000 and the spare is the only maintenance cost – this is effectively identical to the solution in Section C.2.2. Now assume that the spares are kept in an inventory and that the inventory initially has 5 spares in it (purchased for $1000 each at time 0). Assume that when the inventory drops below 3 spares, 5 more replenishment spares are ordered (for $1000 each). Assume that the replenishment spares arrive instantaneously and a zero discount rate. What is the total cost after 20 years?

C.7 Suppose that the time-to-failure distribution used in the simulation in Section C.2.2 was for a particular part in a system and that the part becomes obsolete (non-procurable) at the instant the simulation begins. If you had to make a lifetime buy of parts (see Section 7.5.1.2) to support this system for 20 years, how many would you buy?

C.8 For the example in Section C.2.3, if the component of interest is always as-good-as-new at the start of the year, how many spares are needed?

C.9 For the example in Section C.2.3, if I desire an 80% confidence, how many spares do I need?

C.10 For the example in Section C.2.4, what if there is no preventative maintenance, how many CMs does your simulator predict? Can you verify this with a simple calculation?

C.11 Rework the example in Section C.2.4 with $t_p = 80,000$ miles.

Appendix D

Summary of Notation and Acronyms

The notation (symbols) and acronyms used in each chapter are summarized in this Appendix. Every attempt has been made to make the notation consistent from chapter-to-chapter; however, there are common symbols that have different meanings in different chapters. In particular, indexing and enumerations (e.g., i, j, x) count different quantities depending on the context; time variables (e.g., t, T, τ) all represent some version of time that may correspond to different things depending on the context; and quantities counts (e.g., n, N) appear throughout applied to counting different things.

Chapter 1: Introduction to Sustainment

AVIC	aviation industry corporation of China
BART	bay area rapid transit
CBO	congressional budget office
CFO	chief financial officer
COTS	commercial off the shelf
DoD	department of defense
FPGA	field programmable gate array
FY	fiscal year
GDP	gross domestic product
HSS	health service support
JIT	just-in-time
MRO	maintenance, repair and overhaul
PHM	prognostics and health management

R&D   research and development
RDT&E   research, development, test and evaluation
SME   semiconductor manufacturing equipment
WMATA   Washington metropolitan area transit authority

Chapter 2: The Acquisition of Critical Systems

A_i inherent availability
ANSI   American national standards institute
A_o operational availability
APS   advanced planning and scheduling system
APTA   American public transport association
A-RCI   acoustic rapid COTS insertion
BSM   business system modernization
CDD   capability development document
COTS   commercial off the shelf
DAU   defense acquisition university
DFARS   defense federal acquisition regulation supplement
DLA   defense logistics agency
DoD   department of defense
EADS   European aeronautic defence and space company
EMD   engineering and manufacturing development
ERP   enterprise resource planning
FAR   federal acquisition regulation
FMEA   failure modes and effects analysis
FOC   full operational capacity
FRP   full rate production
GAO   government accountability office
GOTS   government off the shelf
IOC   initial operational capability
ISO   international organization for standardization
IT   information technology
LCC   life-cycle cost
LRIP   low-rate initial production
LUH   light utility helicopter
MDBD   mean distance between delay
MFOP   maintenance free operating period
MOTS   modified off the shelf

MSA	material solution analysis
MTBF	mean time between failure
MTTR	mean time to repair
O&S	operation and support
OSA	open systems architecture
OT&E	operational test and evaluation
PMI	product management institute
R&M	reliability and maintainability
RFP	request for proposals
TDP	technical data package
TOC	total ownership cost
TPS	Toyota production system
WMATA	Washington metropolitan area transit authority

Chapter 3: System Failure

a_s	total number of (software) errors at time $t = 0$
b	failure intensity
β	shape parameter (Weibull distribution)
CAD	computer aided design
C_{Bt}	screening cost per unit time
C_{cw}	cost of resolving a warranty claim
CDF	cumulative distribution function
C_P	unit cost
C_{Total}	total cost
E[ ]	expectation value
ESS	environmental stress screening
η	scale parameter (Weibull distribution)
$f(t)$	PDF, fraction of products failing at time t
$F(t)$	CDF, cumulative failures to time t, unreliability at time t
FIT	failure in time
γ	location parameter (Weibull distribution)
GPS	global positioning system
$h(t)$	hazard rate at time t
HASS	highly accelerated stress screening
λ	failure rate
$m(t)$	failure intensity or renewal density

$M(t)$	renewal function
MCAS	maneuvering characteristics augmentation system
$MTBF$	mean time before failure
$MTTF$	mean time to failure
$MTTR$	mean time to repair
n_u	total units that start screening
$N(t)$	number of (software) errors that remain at time t
$N_f(t)$	number of the N_0 product instances that failed by t
N_{max}	maximum number of repairs of the system (or component) allowed
$N_s(t)$	number of the N_0 product instances that survived to t without failing
N_0	total number of tested product instances or the number of errors at $t = 0$
ω	testing compression factor
PDF	probability distribution function
$POFOD$	probability of failure on demand
$\Pr(\ )$	probability
$R(t)$	reliability at time t
$R(t, T)$	conditional reliability at time $t + T$ given that the product survived up to time T
$ROCOF$	rate of occurrence of failure
SLOC	source lines of code
t, τ	time
t_{bd}	screening time or period
t_0	starting time
T	failure time
T_W	warranty period

Chapter 4: Maintenance – Managing System Failure

ADT	administrative delay time
β	shape parameter (Weibull distribution)
c_f	corrective maintenance cost per event
c_p	preventative maintenance cost per event
CBM	condition-based maintenance
CDF	cumulative distribution function
Cost_{annual}	annual maintenance cost

C_{PHM}	life-cycle cost of the system when managed using a PHM approach
C_u	life-cycle cost of the system when managed using unscheduled maintenance
$\delta()$	delta function
erf( )	error function
η	scale parameter (Weibull distribution)
$f(t), \hat{f}(s)$	PDF in the time and Laplace domains
$F(t)$	unreliability at time t
$FFOP$	failure free operating period
$g(t), \hat{g}(s)$	repair time distribution in the time and Laplace domains
γ	location parameter (Weibull distribution)
$\Gamma( )$	gamma function
GDP	gross domestic product
I_{PHM}	investment in PHM when managing the system using a PHM approach
k	number of spares
λ	failure rate
LDT	logistics delay time
m_a	number of activities
$M(t), \hat{M}(s)$	renewal function in the time and Laplace domains
$\overline{M}$	mean active maintenance time
$M_a(t)$	maintainability
$\overline{M}_{ct}$	mean corrective maintenance time (same as $MTTR$)
MCF	mean cumulative function
MDT	mean maintenance downtime
$MFOP$	maintenance free operating period
$\overline{M}_{pt}$	mean preventative maintenance time
MRP	maintenance recovery period
MSD	mean supply delay
$MTBF$	mean time before failure
$MTBM$	mean time between maintenance
$MTPM$	mean time to perform preventative maintenance
$MTTR$	mean time to repair
μ	mean
μ_{rr}	repair rate (repairs per unit time)
n	number of identical systems in the fleet

n_j	number of jobs
N	number of fielded units, number of available systems, number of independent Bernoulli trials
p	probability of a successful outcome
PDF	probability distribution function
$\Phi()$	cumulative distribution function of the normal distribution
PHM	prognostics and health management
PL	protection level
$\Pr(\ )$	probability
r	discount rate
$R(t)$	reliability at time t
RCM	reliability-centered maintenance
ROI	return on investment
RUL	remaining useful life
s	variable in Laplace domain
σ	standard deviation
σ^2	variance
t, τ	time
t_{MFOP}	MFOP duration
t_p	interval length (preventative maintenance interval)
T	time (actual repair time)
z	z-score

Chapter 5: Availability and Readiness

a_r	number of units under repair (traffic intensity)
$A, A(t)$	availability (generic)
$\overline{A}, \overline{A(t)}$	average availability
$A(\infty)$	steady-state availability
$A(s)$	availability in Laplace domain
A_a	achieved availability
ADT	administrative delay time
A_E	energy-based availability
A_i	inherent availability
A_m	materiel availability
A_o	operational availability
$A_{no\,PHM}$	availability with no PHM included

A_{PHM}	availability with PHM included
A_R	readiness
A_s	supply availability
$A_{sys}, A_{sys-new}$	individual system availability
$\mathbb{A}_t$	state transition probability matrix
ATM	automated teller machine
$C_{\Delta A}$	value of an availability decrease
C_P	individual item cost
D_A	design adequacy
EBO	expected backorders
E_{real}	actual energy generated
$E_{theoretical}$	theoretical maximum energy that could be generated
$f(t), \hat{f}(s)$	PDF in the time and Laplace domains
$f(t_d)$	PDF in the relaxation time
$f(t_m)$	PDF in the maintenance time
F_D	fault detection rate
F_t	failures that need to be repaired per unit per unit time
$g(t), \hat{g}(s)$	repair time distribution in the time and Laplace domains
k	number of spares
l	number of unique repairable items in a system
L	number of systems that can be unavailable
λ	failure rate
LDT	logistics delay time
$m(t)$	renewal density function
$\hat{m}(s)$	Laplace transform of the renewal density function
m_b	number of backorders
$M(t)$	renewal function
$\overline{M}$	mean active maintenance time
$M_a(t)$	maintainability
MDT	mean maintenance downtime
MSD	mean supply delay
$MTBF$	mean time before failure
$MTBM$	mean time between maintenance
$MTPM$	mean time to perform preventative maintenance
$MTTR$	mean time to repair

μ_{rr}	repair rate
n	number of state transitions
N	number of fielded units, number of available systems
N_{AC}	number of systems that have met their design objectives
N_T	number of systems at the state of the mission
OEE	overall equipment effectiveness
p	Markov chain one-step transition probability
p_{ij}	probability that the state is j at T given that it was i at time $T-1$
PHM	prognostics and health management
$\Pr(\ )$	probability
q	Markov chain one-step transition probability
$R(t)$	reliability at time t
s	variable in Laplace domain
S	system
SE	system effectiveness
t, τ	time
t_d	relaxation time
t_m	maintenance time
T	time (actual repair time, or time step for state transitions)
v	percent availability decrease
$w(t), \hat{w}(s)$	time-to-failure distribution in the time and Laplace domains
$X(T)$	status of a system at time T
Z_i	number of instances of item i in each system

Chapter 6: Sustainment Inventory Management

A_o	operational availability
CFM	continuous flow manufacturing
C_h	holding cost per period per part
C_{h2}	holding cost per period per repair part
C_{Or}	cost per order (setup, processing, delivery, receiving, etc.)
C_P	purchase price of the part
C_r	fixed cost per repair batch

C_R	repair cost per part
C_{Totalj}	cost total cost of parts in the jth time period
$D(t)$	demand rate as a function of time
D_j	number of parts needed in period j (demand)
EBO	expected backorders
ECR	efficient consumer response
EFR	efficient foodservice response
EFR	expected fill rate
EOQ	economic order quantity
EOQB	economic order quantity with backorders
EPSLI	end of planned service life inventory
f_{rj}	fraction of repairs that can be made at site j
$I(t)$	inventory quantity as a function of time
J	number of local sites
JIT	just-in-time
k	number of spares
λ	failure rate
m_j	average annual demand for parts at site j
m_0	average annual demand for parts at the depot
METRIC	multi-echelon technique for recoverable item control
μ_d	mean of the demand
μ_j	pipeline, average number of parts under repair (per part demand) at (or being resupplied to) site j
μ_{lt}	mean of the lead time
O_j	order time (ordering and shipping to the site) for site j
$\Pr(\ )$	probability
Q	quantity per order (cycle stock)
Q_r	quantity per order repair batch
QR	efficient foodservice response
Q_s	safety stock
ρ	recovery rate (the fraction of failed parts that can be repaired)
σ_d	standard deviation of the demand
σ_{lt}	standard deviation of the lead time
s_0	depot stock (inventory) level
SL	service level
t	time
T	demand time period, average time for a site repair request to be resolved

TAI	total active inventory
$\theta(t)$	deterioration rate as a function of time
TII	total inactive inventory
T_j	average time per repair for a part at site j
T_o	optimum cycle length
TOAI	total overall asset inventory
T_s	shelf life
T_0	average time per repair for a part at the depot
$u()$	unit step function
z	z-score

Chapter 7: Supply-Chain Management

α	shape parameter (Beta distribution)
AM	additive manufacturing
AME	advanced microcircuit emulation
β	shape parameter (Beta distribution)
C_{after}	cost of buying the part after the design refresh date
CALCE	center for advanced life cycle engineering
C_{before}	cost of buying the part before its obsolescence event
C_{DR}	design refresh cost
C_{DR_0}	design refresh cost in year 0
CEO	chief executive officer
C_h	holding cost per part per year
C_H	total holding cost per part
CLA	component-level authorization
C_{LTB}	cost of a last time buy
CMMC	cybersecurity maturity model certification
C_O	overstock cost
COTS	commercial off the shelf
C_{Total}	total cost for managing obsolescence
C_U	understock cost
D	demand
DMSMS	diminishing manufacturing sources and material shortages
DoD	department of defense
DPA	defense production act

DRP	design refresh planning
E[ ]	expectation value
EOL	end of life
EOQ	economic order quantity
ERAI	electronic resellers association international
$f()$	PDF
$F()$	CDF
FAR	federal acquisition regulation
GEM	generalized emulation of microcircuits
IP	intellectual property
L	total loss
N	total number of parts
OCM	original component manufacturer
OEM	original equipment manufacturer
PC	personal computer
Pr( )	probability
PV	present value
P_0	price of the obsolete part in the year of the last time buy
Q	quantity of parts ordered
Q_i	number of parts needed in year i
Q_{opt}	value of Q that minimizes the total loss
r	discount rate
THAAD	terminal high altitude area defense
U.S.	United States
USAF	United States Air Force
WACC	weighted average cost of capital
WTA	winner-take-all
Y_{EOS}	end of support date
Y_O	year of obsolescence
Y_R	year of the design refresh
ZTA	zero-trust architecture

Chapter 8: System Sustainment Enablers

a	age
a_o	age of the oldest worker in the workforce pool
a_y	age of the youngest working in the workforce pool
A	number of aircraft in the fleet

AC_j^i	administrative cost for relocation of aircraft i to base j (measured in *EFH*)
AF	acceleration factor
AHP	analytic hierarchy process
AI	artificial intelligence
β	shape parameter (Weibull distribution)
B	number of bases where the fleet is assigned
B/C	benefit/cost analysis
BCA	benefit cost analysis
CATOBAR	catapult assisted take-off barrier arrested recovery
CBA	cost-benefit analysis
C_{cc}	cost of new custom code (in personnel hours)
C_{COTS}	direct cost of buying the component
CEA	cost effectiveness analysis
COBAL	common business oriented language
COTS	commercial off the shelf
C_P	acquisition cost per unit
$\overline{CSL_i}$	maximum service life of aircraft i in *EFH*
CTOL	conventional takeoff and landing
CV	aircraft carrier
C_w	life-cycle cost of the system when managed with the change/improvement
C_{wo}	life-cycle cost of the system when managed without the change/improvement
DMSMS	diminishing manufacturing sources and material shortages
DoD	department of defense
EFH	equivalent flight hours
EMI	electromagnetic interference
EOM	end of maintenance
EOR	end of repair
E_n	estimated effort without reuse
E_r	estimated effort with reuse
η	scale parameter (Weibull distribution)
$f(t)$	PDF, fraction of products failing at time t
$f_C()$	current age distribution, PDF
f_{Field}, f_{Test}	system's field and accelerated test failure probability density function
$f_H()$	hiring age distribution, PDF

$f_L()$	exit age distribution, PDF
F	operator fatality rate/mishap
$F(t)$	CDF, cumulative failures to time t, unreliability at time t
FAA	federal aviation administration
F_{Field}, F_{Test}	system's field and accelerated test cumulative failure density function
FHR^i_j	flight hours required for relocation of aircraft i to base j
γ	location parameter (Weibull distribution)
$h(t)$	hazard rate at time t
h_{Field}, h_{Test}	system's field and accelerated test hazard function
H	annual hiring rate
$IEFH_i$	initial EFH of aircraft i
IRR	internal rate of return
I_w	investment in the change/improvement
L	monetary loss (less the aircraft) per mishap
L_{Field}, L_{Test}	system's field and accelerated test life
$Life$	average operational lifetime of a unit
L_{ji}	1 if aircraft i is assigned to base j, otherwise 0
M	number of mission types
MCE	multi-criteria analysis
M_r	mishap rate (per 100,000 flight hours)
$MTBF$	mean time before failure
$n_i(a)$	total number of people in the workforce pool of age a in year i
O	operation and support cost per flying hour
O&S	operation and support
P	workforce pool size
PHM	prognostics and health management
P_0	workforce pool size at time 0
P_1 and P_2	organization-specific parameter used to map the skill level
R	pilot fatality rate per mishap
$R(t)$	reliability at time t
R_{Field}, R_{Test}	system's field and accelerated test reliability
ROI	return on investment
SF_{mj}	severity factor for mission m at base j

S_i	cumulative skill level in the workforce pool
STEM	science, technology, engineering and mathematics
STOVL	short take off and vertical landing
S_0	cumulative skill level in the workforce pool at time 0
t, T	time
t_{Field}, t_{Test}	system's field and accelerated test time to failure
TCE	transaction cost economics
T_0	time to perform the activity with a skills pool having S_0 skill
Δt	technology lag time
UAV	unmanned aerial vehicle
U.S.	United States
USAF	United States air force
VSL	value of a statistical life
WACC	weighted average cost of capital
x^i_{mj}	number of flight hours flown of type m at base b by aircraft a

Chapter 9: Contracting for Sustainment

a	revenue model revenue levels
A	availability
α	customers' share of the contractor's costs of operation
$A_{min}, A_{penalty}, A_{accept}, A_{max}$	revenue model availability levels
A_o	operational availability
BCA	business case analysis
C	cost to the contractor
CfA	contract for availability
C_P	the purchase price of the part
DoD	department of defense
FAR	federal acquisition regulation
FFP	firm fixed-price
FP-EPA	fixed-price with economic price adjustment
GDLS	General Dynamics land systems
GE	General Electric
HIMARS	high mobility artillery rocket system

IP	intellectual property
ISO	international organization for standardization
JSTARS	joint surveillance target attack radar system
k	number of spares
λ	failure rate
$MTBF$	mean time between failure
OBC	outcome-based contract
OEM	original equipment manufacturer
ORR	operational readiness rate
ω	fixed payment
P	contractor's profit
PBL	performance-based logistics
PBSA	performance-based service acquisition
R	reliability
RSP	readiness spares package
R_v	revenue received by the contractor
t	time
TAT	turnaround time
TDP	technical data package
U.S.	United States
v	penalty or reward rate for achieved availability

Chapter 10: Epilogue – The Future of System Sustainment

AI	artificial intelligence
AM	additive manufacturing
CAD	computer aided design
CBM	condition-based maintenance
DCF	discounted cash flow
DMSMS	diminishing manufacturing sources and material shortages
DRP	design refresh planning
ERP	enterprise resource planning
GM	green maintenance
IT	information technology

LCA   life-cycle assessment
MIT   Massachusetts Institute of Technology
PHM   prognostics and health management
RUL   remaining useful life
TDP   technical data package
UV   ultraviolet

Appendix A: Discounted Cash Flow (DCF) Analysis

A annual (per period) value
A_1 base cash flow in period 1
β sensitivity (also called volatility)
CAPM   capital asset pricing model
D debt
DCF   discounted cash flow
E equity
EMRP   equity market risk premium
f inflation rate
F future value
g geometric gradient (per period)
G arithmetic gradient (per period)
n number of time periods
P present value
r discount rate (or real discount rate) or WACC
r_f market (or nominal) discount rate
R_d cost of debt
R_e cost of equity
R_f risk-free interest rate
R_m market return
R_p equity market risk premium
T-Bill   U.S. treasury bill
T_e effective marginal corporate tax rate
V company's total value (equity + debt)
V_{actual} actual value
V_{real} real value
WACC   weighted average cost of capital

Appendix B: Monte Carlo Analysis

a	minimum value
α	confidence
b	maximum value
β	shape parameter (Weibull distribution)
CDF	cumulative distribution function
$\chi^2_{a,\nu}$	chi-square distribution
D_P	Pearson's cumulative test statistic
E_j	expected frequencies (for the jth bin)
Er( )	standard error
η	scale parameter (Weibull distribution)
$f()$	probability density function, PDF
$F()$	cumulative distribution function, CDF
FAST	Fourier amplitude sensitivity test
γ	location parameter (Weibull distribution)
$\Gamma()$	gamma function
LHS	Latin hypercube sample
μ	mean
n	number of samples
n_I	number of intervals
O_j	number of observations in the jth bin
PDF	probability distribution function
$\Phi()$	cumulative distribution function of the normal distribution
P_m	scaled and shifted uniform random number
σ	standard deviation
t, T	time
t_s	sampled time
U, U_m	uniform random number between 0 and 1 inclusive
ν	degrees of freedom
z	the z-score (standard normal statistic)
$\lfloor \; \rfloor$	floor function

Appendix C: Discrete-Event Simulation (DES)

ABM	agent-based modeling
β	shape parameter (Weibull distribution)
CDF	cumulative distribution function

CM   corrective maintenance
$Cost_i$ cost of event i
DES   discrete-event simulation
EOS end of support
η scale parameter (Weibull distribution)
$f(t)$ PDF, fraction of products failing at time t
$F(t)$ CDF, cumulative failures to time t, unreliability at time t
γ location parameter (Weibull distribution)
k number of spares
λ failure rate
$MTBF$ mean time before failure
PM   preventative maintenance
r discount rate (or WACC)
$R(t)$ reliability at time t
t time
t_c cumulative failure time
t_p preventative maintenance interval
WACC   weighted average cost of capital

Index

CPSIA information can be obtained
at www.ICGtesting.com
Printed in the USA
JSHW010832020922
29795JS00002BB/14